T0206056

PRE-ACCIDENT INVESTIGATIONS

# BETTER QUESTIONS

AN APPLIED APPROACH TO OPERATIONAL LEARNING

PRE-ACCIDENT INVESTIGATIONS

# BETTER QUESTIONS

AN APPLIED APPROACH TO OPERATIONAL LEARNING

TODD CONKLIN, PhD.

CRC Press
Taylor & Francis Group
Boca Raton London New York

CRC Press is an imprint of the
Taylor & Francis Group, an **informa** business

CRC Press
Taylor & Francis Group
6000 Broken Sound Parkway NW, Suite 300
Boca Raton, FL 33487-2742



Printed on acid-free paper

International Standard Book Number-13: 978-1-4724-8613-4 (Paperback)



**Visit the Taylor & Francis Web site at**
**http://www.taylorandfrancis.com**

**and the CRC Press Web site at**
**http://www.crcpress.com**

*Humans learn by micro-experimentation.*
*... sometimes those little experiments fail*

Attributed to **Jens Rasmussen**

# Contents

# *Foreword*

Many of us want better answers. Or more answers. We think that our wisdom, and that of our organization, depends on an accumulation of answers. The more answers, the smarter we are. The more we can do. The more our organizations can do.

Todd Conklin, in this book, shows us that it isn't so. Moreover, of course, it never has been so. Scientific progress has always depended not on more answers. Instead, it has been driven by our growing ability to ask ever better questions, to evolve. Ask better questions, and you get better answers. The kinds of answers that will help you and others ask even further questions.

Safety is no different. It is a field in which there aren't necessarily any right or wrong answers in any case. Nevertheless, it is a field in which our questions can surely do with some improvement. Questions like "what rule was broken?" or "who did it?" or "what should the consequences be?" are no longer good questions. They are shortsighted questions that produce shortsighted answers, answers that are not constructive for the safety of your organization, and additionally, are not very good for its humanity and collegiality either, come to think of it.

Todd is one of the best interlocutors in the safety field. You will find his writing accessible, engaging, inviting and refreshing. His experiences with, and in front of groups who want to think about safety differently are wide and deep. It shows in what he has to tell you here, in what he is keen to share. Todd does not pontificate. Todd does not lecture from on-high. Todd invites. Todd beckons you into the conversation. Todd wants you to be part of the exhilarating movement that drives safety differently. Todd wants you to start asking better questions.

**Sidney Dekker, PhD**
*Professor, Director, Safety Science Innovation Lab*
*Griffith University*

# *Preface*

## *Not everything succeeds and not everyone gets it.*

They called him "Captain" and when he came into the training room, he took the place by storm. He had a huge presence. I noticed him because he did not shake hands with his co-workers as he worked around the room greeting people; he hugged them all instead. He actually physically hugged every person he met in the room. Hugging people is odd at a safety meeting in the oil and gas industry. In fact, hugging people at a safety meeting is remarkable. This is a group of workers that have many excellent traits; hugging is not normally one of those traits.

Captain was a safety leader at his company. He was also an excellent worker. He had been around awhile; he knew the work and was a trusted voice in the field. He really cared about his fellow workers safety and well-being. Captain was exactly the type of guy you want in a safety class. I was certain that he would be supportive of the new ideas that were being introduced. Even though I had never met him, I knew he would be a friend on the safety journey.

Perhaps that is why I was so surprised when he got angry. Captain was furious with me. He could barely string his words together, that was how mad he was with me that afternoon. He did everything but stand up and call me an idiot or even something worse. He was upset and unsettled and it was bad. The room got quiet, very quiet.

From the back of the room came this statement, "You have to admit that some people are just stupid. Some people are just not smart. If you tell a worker not to stick his finger into a hole every day, and one day you leave and he sticks his finger in that same hole you warned him about. He cuts his finger off. That guy is just an idiot; he's stupid."

Captain was convinced that he was right on this and I was wrong and uninformed … basically an idiot standing in the front of the classroom. So convinced by this notion, he told me that he could never win an argument with me because every statement he made to me I would always reply, "He was not right."

Then he got up and walked out of the class.

I had angered the one person I had thought would be a sure-fire supporter of my presentation and concepts. I had somehow taken one of the nicest, most sensitive people in the room and made him angry. I had an argument with one of this companies safety leaders, and the argument was in defense of all the stupid workers in this very same company ... for that matter, all the stupid workers in the world.

Captain was convinced that the problem was the worker being stupid. So convinced by this thinking, he was willing to argue publically, in front of all his co-workers, that the world is filled (at least partially) with stupid people that we hire in our companies. If we could only recognize a certain percentage (somewhere around 10% according to Captain's statements that morning) of our workforce is stupid, somehow, that would make all of us safer. So convinced of the stupid worker theory, he was willing to stand up and walk out on this discussion in front of his peers and leadership; so convinced he knew he was right and I was wrong.

I hope Captain realizes the argument I was having with him was about workers just like him. Good, experienced workers who, after some type of accident, appeared to have been injured because they were stupid. It seems strange now, but at the time, this encounter was really stunning to the group ... the belief that somehow a worker who gets hurt must have become momentarily incompetent. The sad part is that these ideas seem to be genuinly held by a majority of workers and managers. For Captain, this belief was so strong that he was willing to argue publically and leave a training class to emphasize his point. Sincerely held beliefs are hard to change. Company cultures are incredibly difficult to evolve. I had hurt Captain's feelings and he certainly affected my feelings as well.

This incident really bothered me. I could not understand why Captain, this safety leader, was so deeply committed to the idea that some workers are just stupid. "You can't fix a stupid worker." Why would workers need to be stupid in the eyes of their bosses, peers, and friends? What more, this idea that the worker is stupid was controlling our discussion in the class that morning. Smart and normal thinking was being derailed by a discussion of how stupid workers were the real problem. We won't ever get better if people choose to remain stupid. Captain was convinced that the people were the problem. The people must be fixed before the workplace could ever be safe.

It also started a new problem for me. I started thinking about the difference between mistakes – unintentional deviations from expected outcomes and choices. A stupid guy purposely makes the wrong choice out of all the possible choices that could have been made. Do workers make decisions or do workers make mistakes? Are workers stupid before the harm happens and smarter after the event occurs? Are decisions choices or are decisions mistakes?

That's the challenge! These deeply held beliefs that there are stupid workers making wrong choices is so compelling and complete of an explanation for an accident that it stops our thinking at that very point. In many ways this type of thinking is so strong it derails all other types of thinking. We stop asking questions or investigating an incident because we don't need to investigate or learn once we have established the worker was simply stupid. I find this to be lazy and incomplete. "I told that worker not to stick his finger in that hole every day for the last three weeks. As soon as I turned my back, the first thing this worker does is stick his finger in the hole and cut his finger." There can only be one explanation for this incident; the worker must be stupid – end of story.

In reality, I have come to believe we have long believed that mistakes are really just bad choices made by stupid workers. Stupid workers don't make mistakes; stupid workers are making choices that smarter workers would not have made. This idea that error is somehow a choice of the stupid is incredibly bad for us. The idea that error is somehow a choice of the stupid is limiting our ability to improve. Mostly what this idea is doing is making us ask really bad questions.

As hard as we strive to help expand the understanding of how events happen in our companies and organizations, we cannot move beyond the decision that the worker made a bad choice. Even our discussions about how events happen are colored by this idea. This thinking takes our response in the wrong direction. This type of thinking is like going the wrong way down a road ... you are moving, but the direction you are moving in is not taking you where you want to go. It is really futile. Wrong thinking costs you and your organization time and money.

Let me offer an example. This example is pretty risky and contentious, but this example is one about which I have spent a considerable amount of time and emotional energy thinking. I am fairly certain that we will not answer this question, as this example is very emotionally charged and socially powerful. Think of this example as a thought experiment around a very important social question. The example begins with one question. Take several moments as you read this example's question to contemplate all that we have talked about in this book. Think about that guy "Captain" and all your leadership team as you ask this question for yourself.

## *Is drunk driving a choice or a mistake?*

If you say that drunk driving is a choice, then you get to blame and punish. You can absolutely hold the drunk driver accountable for their actions. You can arrest and jail the drunk driver to make an example of this person for all of society to see and judge. You can make a strong case that good people choose to never drive drunk. Good drivers must always make good choices.

If you think that drunk driving is a choice, the question you must ask is how can we get drunk drivers to make better choices. The problem with this thinking is that drunks don't make good decisions. For what you gain in accountability, you lose in improvement. You have also anchored all improvement to the least dependable part of the problem – the drunk driver.

If you think drunk driving is a mistake, then you will work diligently to build interlocks and systems that simply "mistake proof" this problem out of the operational environment. You would use breath alcohol machines to lockout this vehicle's ignition systems if any percent of alcohol were detected within the vehicle. You would make it impossible to operate a car with alcohol impairment, and you would be very effective at implementing and managing this control. Cars may be built with this interlock as a standard feature. The solution is stable and dependable; the solution is engineered and reliable in all circumstances.

If you think drunk driving is a mistake, you must, therefore, give up the need (both emotional and societal) to single out and blame the driver for being drunk. You would assume that almost any driver at some time or another could possibly drive drunk. You could not punish or blame the person; you would build a system that would assume that any driver could, for some reason, drive impaired. It would be less important to socially judge the person driving as an evil person. In a way both the driver of the vehicle and the people that might be hurt by this driver are victims of larger system failure; therefore, both of these groups of people would need understanding and compassion.

Whether you think drunk driving is a choice or a mistake is not a question for me to answer for you or for society. I honestly don't know what to think about this question. I hate drunk driving. It is an awful problem that hurts many, many people. There is nothing good or socially important about driving a vehicle drunk, nothing. Yet, it strikes me the solution is to see this problem more as a restorative solution and less of a punishment solution.

That is not the point. That thinking takes us away from the reality of this problem. The reality is what we must manage robustly and effectively. The reality is that we need reform the questions we are asking about this difficult and important topic.

Whether the driver was stupid, sick, or inattentive doesn't really matter – what matters is the consequence is the same no matter what the motivation was for the accident. The consequence of drunk driving is completely agnostic of the cause of drunk driving. The consequence does not care if the driver was stupid or smart, bad or good. The consequence is the same no matter the reason for the failure.

Let me repeat that: The consequence is the same no matter the reason for the failure.

That is exactly where we are on this journey. We want to stop the bad consequences from happening. A broken leg does not care if the worker got the right ladder or stood on a pipe-riser, the consequence is still the same.

Why this thought experiment matters is not the accuracy between and error or a mistake. I am not arguing for a reliable instrument that helps us tell the difference between an error and a choice, between mistake and misconduct, between an inadvertent slip and an intentional violation. As I have gone on this journey, for several decades now, I have started to realize the difference between an error and a bad choice is not very important. In fact, I would look you in the eye and tell you that both errors and mistakes are so normal and predictable – they aren't even interesting … and never causal.

I predict a time on your journey in understanding safety differently, when you won't care if the worker made a mistake or violated a process. In fact, I predict a time when you will think all operational upsets are a result of some type of worker error. We all must learn that we have to begin to develop systems that guard against both error and violation, both choice and mistake, both stupid and smart. We have both. We will always have both. Wishing, punishing, or even training either error or violation away will not work for us; never has and never will. The sooner we realize this important idea the better off our organizational systems and processes will become.

We must get beyond stupid. For a moment "Beyond Stupid" was going to be the name of this book. I was worried that having a book on your desk with that title would reinforce the very belief system we are all working so hard to change. I was pretty certain "Beyond Stupid" as a book title would be either pretty easy or pretty hard to market.

Enjoy this book. Enjoy "Beyond Stupid" as a book written for people like Captain, but also for people like you, your leadership team, and your workers. We want good people to have the best possible ideas and tools to make the workplace safe. We want workers who hug and appreciate each other because they care for each other, not because 10% of them are stupid. This discussion is paramount for our future and the future of safety. This discussion will help us not find smarter workers. This discussion will help us ask better questions to the accomplished and efficient workers we currently have in our organizations. This conversation will make everyone safer.

Let's talk about asking better questions, shall we?

**Todd Conklin**
*Los Alamos National Laboratory*
*Los Alamos, New Mexico*

# *Acknowledgments*

Special thanks go to Sidney Dekker for the Foreword, Kent Whipple for advice and organization, Jeff Segler for the cover design, and Bob Edwards for the case studies ... and to you for being brave enough to dare to question the old ways of thinking in order to introduce new ways of thinking. All of us, you and I, are on a journey toward making the world a better place for organizations and their workers. In doing so, perhaps we can make the world a better place for all.

Keep learning ... do not stop learning. Remember, knowing less does not make you smarter.

# Author

**Todd Conklin** retired as a senior advisor at Los Alamos National Laboratory, Los Alamos, New Mexico, one of the world's foremost research and development laboratories, in the human performance and safety integration program. Dr. Conklin had been working on the human performance program at Los Alamos National Laboratory for the past dozen years of his 25-year career. It is in the fortunate position where he enjoyed the best of both the academic world and the world of safety in practice. Conklin holds a PhD in organizational behavior and communication from the University of New Mexico. He speaks all over the world to executives, groups, and work teams who are interested in better understanding the relationships between the workers in the field and the organization's systems, processes, and programs. He has brought these systems to major corporations around the world. Conklin practices these ideas not only in his own workplace but also in event investigations at other workplaces around the world. Conklin defines safety at his workplace like this: "Safety is the ability for workers to be able to do work in a varying and unpredictable world." Conklin lives in Santa Fe, New Mexico, and thinks that human performance is the most meaningful work he has ever had the opportunity to live and teach.

# *chapter one*

# *Better questions*

> When your process has led you to experiences that you can't understand, you need to ask better questions.
>
> **Betty Sue Flowers**
> *Presence: Human Purpose and the Field of the Future*

## *A new view for safety is appearing everywhere*

It wasn't too long ago that authors like James Reason, David Woods, and Sidney Dekker were researching and writing books that were seen as bold and revolutionary ideas for changing the safety world. These "new safety" books make claims that the events safety people were investigating and reacting to were not the same events that were happening in their organizations. These books made bold claims that changed the world of safety.

These new view ideas were amazing. These authors stated that human error was not a choice and that every investigation that focused on the worker was insufficient in explaining "how" the event took place. More emphasis was placed on event context, local rationality (what the worker was thinking when the event happened that led the worker to believe the next activity would be safe), and the holistic nature of an accident.

Needless to say, these new ideas challenged some very accepted notions of workplace safety, broke some pretty "sacred rules" and strong paradigms about safety management. These new ideas were directly challenging the old school ideas and methods that had served our profession so well for so long. In many ways, the traditional view of safety was beginning to change, and this new change is not the type of change that allows for an easy return to the "old way" of doing safety business. Once these new ideas took hold, there really was no going back. Once safety people realized the power that learning how an event happened was far more valuable than simply fixing who had the problem, the cracks in the old ideas began to form and great rays of sunshine began to appear. In many ways, once we begin to see safety not as an outcome to be managed, but as more of a capacity to be nurtured, there was no way to go back to the old thinking.

*Pre-Accident Investigations* was one of the early, practical guides for understanding this new view approach to safety management. The book did not take a deep dive into the new view, but this book did address a need for a broad-reaching overview of how the new view would look in an organization. This book was written to provide not a deep theory dive, but more of a "safety person's view" of how to begin thinking about how to practice some of these new ideas of safety.

We know that the old "name, blame, shame, and retrain" style of safety leadership has not served us well. We also know the need to move beyond the traditional family of safety compliance is vital to getting our organizations to a new, more effective, level of performance. This journey to "safety differently" is a trip that begins with a solid understanding of human systems and technological work interfaces. The first step of this journey is the realization that human error is never causal, always normal, and somewhat uninteresting. Human error is not a choice. Human error is unintentional; human error is simply a mistake a human worker makes.

We know that people make mistakes. We know that human error is normal. The problem is, in retrospect and loaded with blame, human error seems like a choice. After a bad outcome happens, every move the worker had made seemed like a purposeful and conscious decision to have the failure. The confusion between error and choice causes our organizations to both learn poorly and to create corrective actions poorly. We must be hyper-aware of the potential to react to an event in the wrong way simply because of confusion between error and violation (choice).

As discussed in the first *"Pre-Accident Investigation"* and so many other great new view books, we must be aware of how much we layer on "cause" and "intention" on the top of workers after the failure happened. Most all of our event learning was and is created after the bad things happen. Much of what we have traditionally done in safety is to try over and over again to "fix" our workers after something unwanted has happened.

## *This book is not about traditional safety…*

…at least not the first family of safety thinking that I have spent most of my career working towards and thinking about as a safety professional. This book is not going to be a discussion of safe behaviors, compliance, lost-time metrics, or even handrail, hardhat, or safety glass use. This book will not be a guide to the principles of industrial safety and compliance. This book is not even close to any traditional safety books and that is a purposeful choice. This book is about changing minds and cultures. This book will discuss creating "real" safety in our organizations. This book will consider a different type of safety.

Books about safety are plentiful. Look around your office now; I'd bet you have several safety books within reach. Some of these books are terrific and many of these books are filled with fantastic ideas and passion for keeping workers alive, well, and safe. Most of these books share one idea: "Our job is to ensure to the very best of our abilities that workers go home in the same condition that they came to work." These books ask you to make your workplace safer.

I am not going to ask you to not hurt workers. I don't need to tell you how important it is for workers to not get hurt. There is nothing I can write that is nearly as strong as your inner beliefs around workers safety. You never want an employee to get hurt, ever.

The safety books that we have read and used throughout our careers have been valuable guides and tools for the safety profession with the knowledge that we had at the time. Safety books, and the safety thinkers that have written these books have had a tremendous positive influence over safety programs around the world. I am sure the world does not need another traditional safety book.

There may be enough safety books written. Certainly there are many good books that tell you how to approach safety the way we have always approached safety. The safety world is filled with books that repackage old ideas and traditional methods. However, times have changed and we need to catch up with the times. We need to become wise, not just smart.

It is time for a different kind of book, a book that doesn't spend time talking about safety as a goal to be met and measured. This is more a way to think – a philosophy – a new philosophy used to guide our thinking about how we work. Times have changed; it is time for a book that discusses how we manage our capacity to respond to work in a way that creates wisdom. It is time for a book that does not tell workers how to not get hurt, but discusses safety as a steadily growing body of knowledge about how we work. The time is right to move our thinking (and our libraries) to a new understanding and discussion about safety discovery.

Most safety books seem to reinforce a traditional, antiquated view of workplace safety. This archaic view appears to be built around the idea that workers have the ultimate control over mishaps and failures. Workers must simply be better at working and all failure will disappear. Worker behavior becomes a component of the work that both is vital to understand and control and the key to success. That belief, the belief that behavior is the most significant control for reliable outcomes, is simply no longer adequate in a world filled with technology and complexity. It is a new century and we are still holding on to the last century idea that if you fix the worker behavior in your organization, you will fix safety. We have become wiser in so many ways. We now understand work systems in ways that we could comprehend a decade ago. It is an exciting time to be in safety.

A decade ago, holding workers "accountable for better behaviors" became the watchword our management teams used to get better safety outcomes. Our organizations learned to ask workers to behave better; better behavior indeed will create better results, and better outcomes make a better organization. Like so many quick fixes with the best intentions, whole industries popped up to sell the idea that you could fix all the problems in a workplace by only monitoring and modifying how workers behave.

There have been some rather elaborate schemes created to reinforce the "fix the worker's behavior" approach. It looks like we are managing something other than behavior, but eventually you (and your workers) see through the fancy computer systems, and the slick workbooks to uncover another attempt to make workers better at making behavioral choices while on the floor. In fact, there are so many ways to manage behavior that the competition between safety program sales companies evolved into an enormous competitive industry itself. There are many businesses that want to sell you their "silver bullet" for managing worker behavior. The question has become, "How many times can we read that if the workers and supervisors cared more about their safety, bad things would happen less frequently?" The crazy thing is that these programs don't have the ability to prevent the failures that we still expereience. If this idea worked, it would be working. Sadly, we are still faced with failures and mishaps in our organizations. Our entire society has changed, yet we have been trying to measure the new world with an antiquated formula.

That got me thinking. We have good workers, good supervisors, and good managers, following good safety programs. They are all trying their best to make good decisions and exercise good behaviors. Then why are we not preventing all failures? These are all positive things; exactly what we have been told is the key to preventing failures, yet these fantastic things still allow failure to creep into our organizations.

If the people at all levels of the organization have the right intentions then you must assume the problem is not the people; the problem is the question. If your process has led you to experiences you can't explain, you need to ask better questions. The solution is incredibly simple; we are not asking the right questions to create change in our organization. We have been fixated on safety as an outcome. There is no doubt, we were doing good work with the knowledge that we had; the problem is that we were simply on the wrong side of the equal sign. We were focusing on changing our safety results when perhaps we should have been focused on changing our safety processes.

> In our safety thinking, we drifted to a place where we started asking the wrong questions.
>
> **Senior Leader**

Maybe we don't know what prevents failure? Perhaps we are so uninformed about why failure happens that we have focused on worker behavior because it is easily manageable, controllable, and measurable. Perhaps many of the conditions that lead to failure are not surfacing with our traditional safety and hazard identification tools. If we don't know how to stop an accident, we can't prevent that accident from happening. The how is more important than the why. That is pretty scary, operationally, for our organization. Nevertheless, have no doubt that we can and must improve it.

Which is why this book begins with Betty Sue Flower's words: "If your process has led you to experiences you can't explain, you need to ask better questions." It strikes me that we are not getting the long-term value and change from safety programs, as we know them. I think something must be wrong, if not wrong than misaligned in some way.

## *This book is about wisdom*

> The illiterate of the 21st century will not be those who cannot read and write, but those who cannot learn, unlearn, and relearn.
>
> **Alvin Toffler**

The wisdom that comes from knowing more about the world in which we work, because knowing more tells us about how what happens – happens. Knowledge leads to wisdom and understanding. It is time to recognize that we don't know how to prevent events that were unpredictable and unexpected when they happen in our organizations. We must work to predict those incidents before they happen.

We study the product, not the process; we look at what happened, not what is happening. One of our biggest issues as safety and reliability professionals is that we manage to our organization's outcomes instead of managing and understanding the processes that we use to create those results. It is honestly that simple. That is how we created a whole generation of safety thinking and safety programs that talked ourselves into believing that the only tools that exist to manage safety effectively is to do hazard assessments and manage the behavioral choices of the workers. If we merely paid attention to how our workers behaved, how these workers choose to do safe and unsafe behaviors, we would indeed change the safety outcomes. If we identify every hazard – every risk will be identified and removed, mitigated, or defended. Better choices in the field mean fewer accidents on the books.

Oddly, choice is a dangerous thing to manage as an outcome. Choice is also a remarkable, intricate work component to try to control. I am incredibly confident that we can handle many more parts of our organizational process than just worker behavior. I am so sure that I have bet my entire career on this idea. This is the second book I have written on the topic. And, this is why many of you have brought me into your organizations.

We somehow have determined that the worker choice to get hurt or not get hurt is somehow an option on the menu workers carry with them on the production floor.

How ridiculous is it to determine in retrospect that an accident would not have happened if the worker had made a better choice? How arrogant we have been in assuming that getting hurt was somehow on the radar screen of an operator (or sometimes killed) as an option, a choice. It is as if we somehow believe that workers want to get hurt. I know we know better. I know you know I know. I know that you know this too.

Asking workers to behave better is simply not sufficient to change our safety processes. Behavior change will get your organization slightly different outcomes for a short run, but behavior change alone is not enough to better understand how your organization's processes have let failure opportunistically come forward. We have to amend the question from how did the worker fail the organization to how did the organization fail the workers.

# chapter two

# I hate, "you can't fix stupid!"

> Our systems are too complex to expect merely extraordinary people to perform 100% of the time perfectly 100% of the time. We as leaders must put in systems that support safe practice.
>
> **Jim Conway**
> *Harvard School of Public Health*

I hate the phrase, "you can't fix stupid." It is offensive and mean. Most importantly, that phrase is just wrong. Stop saying it. Stop using this phrase right now. "You can't fix stupid" is serving you and your organization poorly. It colors your thinking; it makes you stop investigations too early; it sounds like something a "jerk" would say, and it pisses people off. "You can't fix stupid" causes a worker who already feels bad to feel much, much worse.

I think when people say, "you can't fix stupid" they don't intend to be mean-spirited; my guess is the person who says these words intends to be clever and funny. It is not funny or amusing, and worse yet it communicates the wrong message. It causes managers and supervisor to approach an operational problem with a bias that the reason something bad happened is because the worker was just an idiot. Wrong thinking equals wrong solutions; wrong solutions equal bad outcomes and wasted resources.

When has name calling ever created a safer organization? Wishing people were smarter is not a good safety management strategy. Hoping bad things won't happen is really a weak management position from which to lead. And, frankly, it just hasn't worked … at all.

I would add to this discussion a further complication; your workers are often so responsible and feel so much pride in what they do for your organization that they believe that in order for a bad thing to happen, they had to have been stupid. Good workers are very dedicated to and responsible for your organization's successes. Good workers take pride is doing the right things. When something bad happens, it must go without saying that a good worker would step up to the plate and admit momentary stupidity. You don't have to do anything to further perpetuate the idea that you can't fix stupid.

## *People don't just become stupid*

We must battle the need to believe that when a worker has some type of bad outcome, that adverse outcome happened because the worker became momentarily incompetent. Our thinking is driven by a bias towards bad things happening because someone did something bad. The bias that worker became stupid is really strong force in how we learn from events. Thinking that the worker is stupid sends you down the wrong road for operational discovery.

If the worker was doing an excellent job on Tuesday, what changed so drastically so that this good worker now becomes a stupid worker on Wednesday? If the employee was meeting production goals, behaving helpful to other people, and not being injured, was that worker stupid before the event happened? When we call workers stupid are we not just pushing blame away from ourselves and from the organization we represent. We tend to treat workers after a mishap happens as if the workers have become defective. That belief that the worker is bad or defective must be the origin of the phrase: "You can't fix stupid." It also implies that you were dumb enough to hire an idiot. Either that or managers and safety people are mean, evil, and lazy. You and I know better.

That phrase makes me sad and angry. When I respond to a person who has just proclaimed for the entire benefit of the room that stupid can't be fixed, I want to scream at the individual who said those words.

Screaming won't help either. When a person responds as if the worker has become stupid, that response is more about the organization than the worker in question.

I have had to learn that my reaction cannot be one of anger or aggressiveness; my response must be to teach. It should be all of our answer to this. We all have teachable moments and opportunities on a daily basis. It is essential that we help managers and leaders realize that saying stupid cannot be fixed is not funny or amusing; it is dangerous and colors our thought process and reduces the way we learn. We are smarter than mere name-calling and blame.

We must have this conversation every time someone utters, "You can't fix stupid." For the most part, none of your workers are stupid and neither are you. Your workforce is made up of people who simply want to do the work you need to get done. In fact, the workers you hire carve out part of their own self-image by doing a good job for you. You don't hire stupid workers. If you do hire stupid workers, you are perhaps the worst manager ever. Only a stupid manager would hire a stupid worker. Enough name calling ... enough.

It is time to start teaching our leaders and managers that there is a new way, a better way, to frame safety and the understanding of operational failure. We must not stand and listen to a litany of reasons why the worker should have done something different. We already know it didn't work. Thinking about what the worker should have done in retrospect has no value and represents no learning. It is time to learn better about how work is done. It is time for better questions.

Think about this from one of my favorite thinkers about the reliable performance of smart workers. In this instance, Jim Conway was talking about doctors and nurses in a paramount and significant hospital in the United States. Jim was concerned significantly over the way healthcare as an institution was moving towards unacceptable levels of "perfection expectation" from its very dedicated workers. Read this quote again and tell me does it not apply to your people and the expectations you place on those people?

> Our systems are too complex to expect merely extraordinary people to perform 100% of the time perfectly 100% of the time. We as leaders must put in systems that support safe practice.
>
> **Jim Conway**
> *Harvard School of Public Health*

# chapter three

# To ask better questions, first understand and stop blame

A phone rings…

> Hi. I'm John Smith from the plant. I need your help. We had a pretty serious accident at our plant and we need to know what we should be doing next. One of our best guys was trying to get a belt roller aligned with a machine and he got his hands in-between the rollers and the belt. He got hurt pretty badly. We stood down the plant, sent all our workers to belt safety training, we rewrote our lock and tag procedure and asked them to be safer. What should we do next? Is that enough?

You know about these calls. You have perhaps made or received a call similar to this one. What is so odd about these calls is that they all seem to begin the same way; a good worker gets into some kind of situation that creates an adverse outcome. These calls have gotten to the point where I can almost predict what will be said to me during the call. Some plant manager from a factory somewhere will introduce himself to me and tell me that there has been an accident. I almost never can predict what the accident was, what happened. Honestly, the more of these types of calls I receive, the more I realize that what happened, the consequence of the event that spawned the phone call doesn't matter very much. A bad thing happened and an excellent worker got hurt; that is sad and terrible. I don't know what, where or how the accident occurred at the plant in question, but it is a pretty good bet I will know the type of worker who had the accident. Not the person's name or specific identification, but I can always tell you how much experience the worker has or how good the person is or how much cultural and safety leadership the person offers the facility.

Most of these calls describe a worker who has the following characteristics: a significant amount of years of experience; One of the best workers at the plant; A safety leader in the facility; A peer leader and problem-solver for the company. All in all a good, experienced, and smart worker somehow had an accident. A good person, in a good company, has had a bad outcome.

The plant manager is concerned and confused by this accident. How could one of the best, most reliable workers at a facility have an accident? Nobody can understand how this could have happened to this certain worker. The management of that organization seems to believe for some reason this good person had a moment of stupidity and ended up getting hurt. In so many ways, the belief that a good worker has somehow become a bad worker is the only line of reasoning that could possibly make any sense to the manager's understanding of how his or her organization is, was and will in the future operate.

Our brains don't operate well after a failure; the emotional need to understand "why" the bad thing happens confuses our technical need to know "how" the event transpired. We have to figure out what happened so that we can assure ourselves that this accident was some kind of fluke. Promise me that this won't happen again, not on my watch.

These stories are sad, simple, and dangerous. The desperate need to be able to find a reason why a good worker could get hurt is a strong organizational force and it is not a force for good. It forces us to create simple reasons for how bad things happen. We look, in retrospect, at the terrible things that have happened to find a place where something bad has happened, a place where a worker made a wrong choice. Once we determine that the worker lacked competence and failed, we know that the problem belongs to the person who failed.

If you have studied the "new view" of safety at any depth, you already know that workers are an easy target for blame. Every time somebody gets hurt in an accident in our organization – something wrong happened. The fact that a worker did something wrong is not a surprise; it's an incredibly short-sided way to think about the particular failure, but it is not surprising. The problem is that type of thinking causes us to dutifully fix the workers that follow the injured worker who failed. We are shutting the barn doors after the cows have gotten out.

As discussed earlier, an excellent worker does not become a stupid worker in the blink of an eye. If this person was a "go to" guy on Wednesday, what changed to make this person a liability to the good order and safety of your organization on Thursday? The answer is: Nothing changed with the worker. The worker, most likely, is only trying to make work happen in the midst of an organizational system. Something in the plant that either was not working well or at the very least was working differently at the time.

When workers learn to use your plant's systems, these workers are learning what it takes to do work at your factory. Workers learn to adapt. Workers learn how to make work happen. Those same adaptations and creative risk-taking actions that the operator uses every day to do their job make the worker a success story when work goes well. When work goes badly, all those tricks and habits are bad, at-risk behaviors.

Nothing changed except the outcome. The same behaviors that make workers successful cause workers to fail. The worker didn't fail the company's systems. The company's system often fails your best workers.

It perplexes me that leaders of organizations jump immediately to the idea that the worker somehow had a moment of stupidity. It bothers me that the worker will feel deep in their psyche that he or she had a moment of stupidity that lead to this accident as well. The worker will often lead the "I am stupid and you can't fix me" yell. It all seems to come down to blaming somebody because something did go well. We love to blame. Finding fault somehow seems important, but blame isn't actually that important at all. Blame is a waste of the organization's resources, time, and everyone's energy. None of these things are complete explanations of how the accident happened. But just knowing who screwed up is so compelling, so complete, and is strong enough to stop asking more questions about the event, about the worker, about the facility and about the systems. Your organization has an answer to why something bad happened. If you have the answer to the question, it is time to stop looking around for people to blame and start fixing the problem. Fixing is more fun than learning. Fixing is more rewarding than learning. Fixing gets something accomplished, or so it seems, whereas learning seems actionless as if you are not doing anything.

There seems to be a conflict between the knowing who did the bad thing that led to the event being so compelling of an explanation that more information is not necessary. The more information only seems to reinforce the failure, appears to be the worker's fault, or is it that we don't have the time or the ability to look deeper into the event to try and understand more? We seem to be confusing "who" did it with "how" it happened. The "who" and "how" are very different concepts and will lead to very different operational outcomes. Do we stop looking because we can't handle the extent of the answer we might uncover? Not knowing whom to blame seems pretty scary to us.

It seems so sensible to look deeper into every event – to see the complex context of a failure and "deep dive" learn how such a thing could happen, but that doesn't seem to happen. Not knowing is not scary; not knowing is the only way to get to the deeper understanding that is required to both explain what happened, and more importantly, to fix the problem. And, most importantly, to keep it from happening again.

Which brings me to other side of this struggle, and that other side is the fact that we think a good worker has become temporarily insane. We have enough information to make the story complete; we now know and understand how the story of the accident ends. If that good worker had somehow not become defective, somehow become incompetent, we would have not had an accident. We must stop defective workers from hurting themselves before they hurt themselves.

The problem is that this "blame and fix strategy," may feel as if telling the story of how a worker became temporarily defective seems complete. This story simply is not complete enough for our organizations to learn from, to understand and explain the event, and most importantly for our organizations to fix our workplaces so that we can get better, so that we can avoid having this accident again.

I guess the question that all of this begs is this: "Good managers, managing good workers – all of these folks wanting to do the right things – are satisfied with explaining an adverse outcome by convincing each other that this bad thing happened because the good worker has somehow become a bad worker, at least for the moment that the bad outcome occurred?" Remember, good workers don't have accidents ... or at least that is what we believe before bad things happen.

That question seems ridiculously complicated and arrogantly simple at the same time. "How did a good guy have this accident" is unacceptable, overly simple, and an incomplete way to think about a failure. Yet, I must see some form of this question asked by really smart, good people at every accident that I investigate.

This is incredibly serious. It is almost hard-coded into our investigation language.

Worst of all, the question of how the worker messed up, borders on malpractice. Why malpractice? With this mindset you will make significant operational decisions about the future of your organization with bare minimum of information. And, if that is not bad enough the scant amounts of information you are about to use is almost always wrong. You are looking at the problem wrong, and because you are looking at the problem wrong, you are getting flawed conclusions.

You would no sooner make an operational decision about your production work based on this very narrow, very incomplete view of your organization than you would eat your hat. When you make technical decisions, you gather vast amounts of information. You assume that nothing is perfect and you are incredibly disciplined and checking your work twice for accuracy and completeness. Your organization is simply too complex and too valuable to make significant operational changes based upon an incomplete understanding of what happened; the same idea that Jim Conway talked about several pages before in this book. When you must make an important operational decision, you collect information from many sources and many places. You want the most information you can possibly gather so that you are in a position in which you can make the best possible decision to ensure the best possible outcome. We don't like to make an operational decision based upon partial information. We work hard to have as much information as we can possibly have when we make production decisions.

You wouldn't buy a house this way. Better yet, you wouldn't buy a car that way. Purchasing a car seems much less threatening and easier to use as an example. When you buy a car you will research, discuss, debate, ask around, and think deeply about that type of significant purchase. If it is a car you are going to have a while, you must be as certain as you can be about that decision before you make such a commitment. Why would you explain a failure in your organizational operations with less information than you would have when you buy a car? We must not stop information about events. We must improve the way we gather information about the operational failure.

We have allowed our investigations and event learning activities to become prisoners of placing blame and accountability. In our need to understand what the worker did wrong, we have developed tools and expectations that simply reinforce the discussion of how the worker screwed up. Those tools cause our investigations to stop too early and bias our thinking towards the idea that we must fix the workers. This process doesn't halt the event from happening again. Perhaps even more dangerous is the fact that not only do our investigations and learning activities reinforce the old view of safety, our thinking and problem-solving tools no longer help us learn. They still don't solve the problem and will allow it to happen again. Our thinking and problem-solving tools only seem to serve to reinforce blame.

Before we can become better at learning, we must first understand how important it is to think about failure differently. We must first realize that our tools for investigating are forming our thoughts about failure. If we desire different thoughts, we must find different tools.

We used to solve complex medical problems with blood-letting. We thought we knew what was best based on the information that we had at the time. However, we evolved, we learned new things; we took that knowledge and became wiser. Look at how the medical field has evolved in the last decade because the physicians started asking better questions, developed new tools and created better systems.

# *chapter four*

# *Access knowledge from the field and the floor*

I was thinking...

> We all see our organizations differently, yet we are all looking at the same organization. The bosses see the organization as a place that makes money. The engineer sees the organization as a series of process-flow diagrams that create a product. The worker sees the organization as their livelihood.
>
> Everybody is looking at the same place and seeing things differently.

## *The scene*

Five workers are standing around a flip chart set up on one edge of a receiving dock. These workers are carefully listing almost every activity that they do nearly every morning in the order that these tasks happen. The list begins from the moments the workers arrive and proceeds to document the little "first of the day tasks" that are done each morning. A general overview of what happens each day is collected from these five workers.

This list goes in to a moderate level of detail. The list is more complete than a schedule, but not as complete as a formal job-task analysis. These workers are basically writing down what they do when they start their day at work in the receiving department of their factory. They are creating this list as a group, with discussion and disagreement until they are convinced that they have a relatively accurate account of how work happens to them.

The workers are animated and involved. The workers are laughing and talking, even sometimes disagreeing with each other. Within the first 45 minutes of creating this list, the workers have filled nine separate pages of flip chart paper. Each of these pages has been stuck to the boxes surrounding the workers with fat, blue masking tape. Most likely for the first time ever, these workers have created an agenda of all the activities that they must do to start a typical day in receiving. More improtanly, they did it on the work floor as a group.

Before long, the five workers have an excellent understanding of how work happens in this particular receiving area. These workers could understand and talk about the relationships that different tasks had to all the other tasks that must be done (are being done) every day. It was easy to see conflicts in production and planning that exist on a daily basis that these workers handle not as some type of a special case but as a regular part of their day.

This list helped better understand and record something that could not be represented on a timeline. The fact that these multiple tasks happen not in any particular order, but every morning these tasks occur in the same order, day in and day out, as a part of doing this group's work is not a part of how work is done by these workers – it is the reality of the work. The work that must be done is just that – the work that must be done. The complexities of starting work on a receiving dock are recorded on this list of activities.

You would never believe that these same workers had a near-fatal incident involving a fork truck and a pedestrian just the day before. And that this activity is going to be used in the investigation, no strike that, the event learning.

Standing around a flip chart certainly doesn't look or act like a traditional investigation. In fact, these workers seemed much more as if they were participating in a team-building activity rather than some type of a response to a near-accident. How could listing the things you do every morning have any value at all for a root cause analysis or an event timeline, or a cause and effect chart?

There are no witness statements. There are no individual interviews. There is no finger pointing. In fact, this group did not even speak about the near-fatal incident that had happened the morning before. This group was asked to not focus their discussion on the incident, but instead to help create a complete list of how work is done. Starting from the beginning of the day and working toward the incident, which is very different from a traditional investigation that would start from the event and work backward towards the start of the day.

Suddenly these pages of the flip charts become a way of communicating and understanding how this group functions. Where are the safe and stable parts of the work? Where are the complex operational conflicts that exist while doing this work? Where are the workers at their best during this work and where are they at their highest potential peril?

## *The problem*

Soon the group had a complete understanding of what had happened the day before. These workers recognized that three important activities for the plant they support all occur at the same time each and every morning:

1. The individual plant areas bring their own fork trucks to receiving to pick up their daily receiving, which means there is additional vehicle traffic in a tiny area around 9:30 am each morning.

2. The workers have their first break at 9:30 am each morning and the bathrooms and break area for the plant are located next to the receiving area. This requires the workers walk through the receiving area to get to the restrooms.
3. The external deliveries, of which there are three daily, all arrive about 9:00 am. The dock's three dock doors are all filled to capacity and all the receiving workers are busy unloading these daily loads.

The near miss incident happened at about 9:45 am. Any questions about what was happening that morning in their receiving area?

Interestingly, had a traditional approach been used to understand this incident the same information could have been discovered by a team of safety professionals. My guess is if it were identified it would not have been seen as sufficient enough to be the cause of the pedestrian strike. I would bet my paycheck that the investigation would have led to the driver of the fork truck that hit the walking worker. That driver would have been corrected in some way to communicate to that driver (and all the other drivers) that our organization does not strike walking workers with moving vehicles.

That corrective action might have worked. As corrective measures go, it certainly seems like the one that would be predicted to be used in most facilities. After all, if the fork truck driver would have seen the worker and yielded, the worker would have not been struck and injured by the fork truck. Blaming the fork truck driver is an answer for this operational failure; it is just not a very complete answer.

## *Blaming the driver is not a long-term solution*

By engaging the workers in problem discovery and problem solution, the company was able to both motivate and understand how work processes and systems in this receiving area had over time created the possibility for a fork truck to strike another person. The components and conditions for this incident were in this receiving area every day. The operational conflicts of scheduling everything to happen at one time were present every day. The presence of lift vehicles from both receiving and the plant floor is how the plant keeps production flowing, and the bathrooms and break room have been located in that part of the factory for almost 100 years.

## *The learning*

Perhaps a better question would be, "how did the receiving drivers not strike other workers every day?" As this team of workers completed the learning activity, the depth of discovery and wisdom both the receiving workers and the plant management gained from their operations was so profound that the corrections for this problem were simple. This learning activity identified three operational conditions that needed to be moved, rescheduled or formalized.

1. Make a parking lot for plant fork trucks to park outside of the receiving area. The worker took this action and performed a small experiment using the fat, blue tape to mark off an area outside of receiving to be used as a waiting and load transfer area. If this marking worked, the plant would formalize the area with paint and signs. If this method did not work, the workers would pull up the old tape and put down a new parking and loading configuration.
2. Put in a green safe to walk – walkway around the outside edge of receiving leading to the bathrooms and break area. Again, the workers already had moved racks and cases away from the wall of the receiving area to create space for a walkway around the parameter of the outside wall of receiving. Management had already begun securing railing and preparing the floor to be painted green and yellow.
3. Stagger the schedule for the outside deliveries at different times during the day. By the end of the first 45 minute Learning Team meeting, the plant manager had called all three delivery companies and rescheduled the trucks for times that the workers felt would be better for each delivery – some earlier and some later.

What about the driver? Shouldn't something be done to make that driver aware that he had come as close to killing a person as you can get without actually killing the person? Don't we need to hold that operator accountable for this near miss? If we don't make a big deal out of this doesn't it mean that I am not managing my workers? Will we have anarchy among our fork truck drivers?

What this management team did with this driver was to provide support and to console him. The plant manager apologized for the environment that management had given him to do his job. He had a very significant emotional and operational event. He was nervous and sad and worried and, most significantly quiet. He was certain he would lose his job. The last management action he expected was to find understanding and support. That driver is now a safety lead for fork truck operations in this plant. He is smart, wise and concerned and involved in the safe operation of fork trucks plant wide. He was not fired; he was engaged and motivated to lead his peers in being world class in the movement of equipment, parts, and supplies to his plant. This is not a radical concept.

And the plant got better. Receiving started to be very efficient. Overtime for the receiving works has all but gone away. The plant moves better, and receiving, once a giant bottleneck for the production, is now fast, clean, and just in time. The problems were worker- discovered and the solutions were work-owned. These workers are engaged in finding operational complexities. They make them visible to management, and provide solutions that are sustainable and long-term for these operational conflicts and issues; but, that is not the best part of this story.

The best part of this story is that the management team got much, much wiser. When we give leaders enough information, information that is complete, detailed, and tells the real story of the failure, something amazing happens. Managers make wiser decisions about their operations. Better inputs always equal better outputs. The only requirement that these workers and managers had to address was how brave they were going to be. Giving workers a voice in problem identification and problem solution means that the manager must abandon some control. Managers must be courageous enough to realize that the only people who know how the work is done are the people who actually do the job.

# *chapter five*

# *Not knowing is powerful*

> The skill is not in knowing the right answer. Right answers are pretty easy.
>
> The skill is in asking the right question. The question is everything.

This book is about change. Not changing the way you manage safety or production. Not changing people or procedures. This book is about changing the questions you ask about reliable and stable operations within your organization. You don't know what you don't know about your operations and that should scare you a little.

We are not excellent learners. Let me take that back, we are actually quite good at learning; we just do not learn well from failure when it happens in our organizations. The reason we don't learn well is that we don't ask the right people the right questions. When you ask bad questions, you get bad answers. Our questions seem to all be directed at reinforcing the notion that the worker was momentarily incompetent. What is even more interesting is that we often don't learn on purpose; quite the contrary, most of the best operational learning done in your organization is done by accident. The lack of purposeful learning is really a problem operationally.

> Since this is about measuring intelligence, which could be useful in an interview scenario, then I'd say that "asking the right questions" is actually measurable whereas "knowing the right answer" is not.
>
> **Fahd Butt**

In reality, organizations don't learn at all. People learn within organizations. Organizations don't really get better. Operations get better because workers evolve, systems evolve and organizations do perform more efficiently and safer. This individual improvement makes for better organizational outcomes and organizational memory. Organizations don't change well either – people change within the organization and, in turn, the organization becomes different because the people are different. People learn, and because people learn these same people also know more. The workers evolve. These evolved workers are more efficient and more productive. Soon the organization is performing well because the people in the organization are performing well. Seems pretty basic, but in many ways the idea that an organization is a group of people has been lost over the years. Organizations are you.

As the organization and its people get better at doing the things that the organization does, operational information (how work is done around here) becomes "hard-wired" into the organization. With repetition, the operational information becomes the standard way work gets done. This information is crucial; you count on this information more than you know and become the operational memory for the organization.

Institutional memory actually comes from the people. The organization has no real memory. The organization has no real culture. All of those things exist in the people that make up your organization.

We spend much of our time determining what we think is true from what we believe is not true. Our job, as people who have taken on the responsibility for the safety and productivity of workers within our organization, is to learn what is happening – what is dependable, real and predictable. We seek the truth, the operational truth. What we must know is "how" is what is happening in our organization, happening?

The problem is that within our understanding of our organizations, what we often think of as true is not true … and what we think of as not true is often true. There is only one way to differentiate between these two, true and false. We must go to where the truth is made. We must go out into the field and find out. We must ask questions, lots of them. We must talk to the people who do the work. We must observe our organizations. We must learn. What must identify the way work happens when workers do work? What conditions surround the workers while they are doing work? Knowing what conditions exist helps us to better understand the context of how work is done. Workers are always managing work conditions. Many of these conditions are not known to us but are very apparent to the workers.

Once conditions are identified and understood, then the process of learning how the event happened can begin in earnest for the organization. Our job is to help the worker tell the story of how work happens in both success and, unfortunately, in failure. It's about questions, not timelines.

Our current tools for managing safety seem to emphasize the worker over the work conditions. We have almost gotten to a point where we investigate failure to find not how the failure happened but to identify why the failure occurred. The failure occurred because something or someone failed – that is not even very important to understanding the story. I am not sure organizations even know why they do investigations anymore. We must know why we do event learning for us to ever improve event learning.

If you take nothing else out of this discussion, be able to answer this question: "Why do you do event reviews, investigations, and critiques?" Are investigations done to blame, punish, and establish accountability, or does your organization do investigations to learn and improve your organization? This book hopes to reframe the many and significant answers to this question.

Investigations are done solely to learn and improve. Blame, accountability, culpability, and fault are outcomes that exist after the learning has happened. These things are never a reason that we learn and they never should be the reason we learn.

But, thinking differently and doing different types of investigations is not easy. The craziest thing about the questioning of why an organization does investigations is that I find large, well managed, famous, organizations often have a very difficult time coming up with an answer to that question.

The only reason you try to better understand events is to learn and improve.

Have you ever had the "A-ha" moment, the moment when, suddenly something changed? You discover all you have learned and trusted is flipped on its side. This "a-ha" moment typically happens within seconds. Nothing actually happened within your environment. Everything that was there was there. You changed because you became wiser. You saw your world from a different perspective. You learned that something is different. Not that the difference was not there all along. It is always hard to notice something that doesn't happen. All that you knew before was okay? However, you became smarter.

> The test of a first-rate intelligence is the ability to hold two opposed ideas in mind at the same time and still retain the ability to function.
>
> **F. Scott Fitzgerald**

That is why I wanted to write this book. We don't spend nearly enough time learning about our organizations. What we think of as true is often not true. What we believe is clear may be opaque. What we think of as simple to do is often hard to do. Rules that seem complete are usually incomplete. What we think we know may not be real, but because we believe all these things are true – we have no reason to question and understand what we think. The enemy of the question is always the answer. If I know something, then I don't need to learn anything.

There is good news!

Not knowing what is true or untrue, what is real or unreal is actually kind of interesting. Not knowing gives you freedom. Not knowing allows you to go out into the world to discover and learn – without bias – with understanding and a desire to find out more. This truth or non-truth search is nothing more than an opportunity to learn what is happening based on how we look for what we find. If you change the way you look at performance, the things you see change.

## *Safety is about learning*

Safety is a practice. Safety is a process. Safe and reliable performance is a process to be guided, not an outcome to be managed. Safety is wisdom. We create safety, or more importantly, we co-create safety in the relationship between workers, planners, managers, and the tasks being accomplished in real time. None of those components mentioned above involves the scoring and tracking of work outcomes. What matters are the real-time relationships each group has with the other group? We can't manage the results these relationships create because the outcomes are a result of those relationships. We have to become better at understanding the perspectives and relationships that these groups have with and about one another. Reliability is found not in the individual workers and managers, but in the relationship between the workers and the managers.

We no longer create safe and reliable workplaces for workers. We now know that we co-create safety and reliability with the workers, your workers produce outcomes often in spite of management, not because of management's brilliant guidance and direction. In reality, we must be productive together, in co-creation of work success.

You see this all the time. Where once you selected a restaurant because of its decoration, its billboards, or its advertising efforts, you are now much more likely to read a review on Yelp or Trip Advisor. You make your decisions based on feedback from other consumers. The world now co-creates knowledge. Reputations and businesses are made and lost based upon the collective information created about an organization. This same idea must be true for your organization. New century, new methods, change is upon us and we must respond.

## *Learning is a product of feedback*

Understanding what is happening while it is happening is the primary requirement for successful learning to occur. Near as I can tell the only way to guide a process, is to be regularly pulsing that process to understand, and manage where that process is heading. That is the definition of a feedback system. Build a communication culture that allows you to talk to each other in real-time about good news and bad news, problems and solutions, and processes that are known and unknown and allow that system to function. That's the operational idea of a feedback loop: constant monitoring of a process, while the process is stable and functional – feeding back into the system itself. All of the while this feedback is done while successful work is being done, normal and stable performance for your organization.

This feedback loop, the ability to monitor a process while the process is in practice, is an essential learning activity. In fact, the feedback loop is vital to how humans develop. Yet, we tend to measure our safety processes by what happened, when we should be more concerned with what is *happening*. We see safety as a thing to accomplish, not a practice to be applied and in doing so we limit the amount of feedback we collect about the process while it is happening.

What our organizations are doing by hyper-focusing on operational outcomes is restricting our ability to learn about our process by focusing on the end product as opposed to the countless complexities that create the path towards success and failures. When we concentrate on the outcome, we miss the process. Yet, when we concentrate on the process we absolutely are left with a much better, much more complete understanding of the outcomes. This is somewhat mind-blowing in a way, but don't let that freak you out, because it is also that simple. We know that we want predictive information – so therefore we have to start looking at work, as done, not the work when it is done.

## Outcomes matter

There is no doubt that outcomes matter. They do matter. Many of you and probably all the individual members of your executive leadership team have your personal and professional worth determined by the outcomes of your safety and reliability programs. And, that was good until now. The incentive structure that we use to both understand success and score our company's place in industry comparison statistics, the world we exist in, is built around these outcome measurements. These outcomes are so strong and so significant that they have colored the way we see the world and the way we do our work.

The question we struggle with is, do these outcome measures matter so much they distract us from looking at the things in our organizations that could make the biggest difference to our safety and reliability programs? My answer to that question is giant "yes." We are so attracted to the idea of having no failures that we have become blind to the fact that failure is normal and an important component of learning and improving. It is evolution. It is the reason pencils have erasers and computers have a delete key. Organizations have lots of failures. For this reason, because of our very high reliance on outcomes to determine programmatic success or failure, we are not very good at understanding how our processes affect our organization's safe and reliable performance. We have historically allowed our organization's outcomes to ultimately determine our organizational processes and systems.

Almost all evolution comes through failure. We tend to learn more from the everyday things that go wrong. The reasoning behind that has to do with understanding the change in our outcomes is a product of a change in our processes. None of us loses our passport twice. If you lose one passport and have to replace it in the middle of nowhere, you quickly change your process for storing your passport.

Yet, without a measurable outcome you are sunk. Your organization exists to perform some function. Without that function, there would be no profits, no production. Without profits, there would be no organization. All organizations are in place to perform some function and that function is important to measure and compare. We can use the outcome to draw comparisons between teams, between locations, between the border of countries, between this year and last year. Outcome numbers are easy to manage and easy to calculate. Outcome numbers are so quantitative that these numbers naturally devalue the scariest cousins of outcome numbers, the non-measurable qualitative information. We pay attention to what we can measure. We tend to collectively ignore, either consciously or unconsciously, the things that we cannot measure. We favor what we perceive as non-emotional, fact-based, measurable numbers, not what we think of as the weak and subjective stories about how people think work happens. Sometimes quantitative measures are not encompassing enough to accurately describe the more nuanced qualitative realities that exist at the work face.

So, there, you have it. We measure what we think matters and our incentive structures support the idea that outcomes matter most, which strongly reinforces the need to measure what we measure most accurately … even though we know that outcomes are the result of the processes and systems that we use to perform work. The problem is that we can't stop measuring outcomes because we are on the hook for having the outcomes of our organizations measured. In some ways, we are caught in a trap of measuring and scoring metrics that matter on the product delivery side and probably cause our considerable misdirection on the production delivery side of the house.

I want to make the case that our organizations have somehow, over time, separated safety from operational learning and it is not a good thing. In fact, it is a dangerous thing. That separation causes good leaders to think about events in the wrong way. It punishes people and it actually hinders the learning process. We are holding ourselves back with the best of intentions.

There is a profound difference between measuring and understanding *outcomes* versus measuring and understanding *processes* of doing work. Think of this problem in as simple terms as the difference between system outcomes and system performance. In our zeal to produce outcomes, we have moved away from the actual act of production. In our zeal to operate, we have moved away from our operations. In our zeal to know the answers, we seem to have forgotten how to ask the questions. Our need to get answers has stopped our ability to ask questions. We have made our systems so production-centric; our systems no longer have the capacity to allow us the space to learn.

## *And learning matters*

We have to intervene in the cycle of outcomes as a meaningful way to understand the practice of safety and reliability. We have to move our operational focus from outcome-based management to process learning management. In many respects, we are invested in a system that leads us to concentrate more on the action without learning instead of action because of learning. We must create a world where action is the result of learning. Getting better at operational learning is the only way we can move safety and reliability to the next logical place.

Learning is not the same thing as knowing. There are volumes of theories written about the importance of learning. There are plenty theories about how people learn. These foundations are necessary, but not the topic of our discussion. We want to talk about our ability to create space within our operations for the art of discovering information, the collection of operational intelligence. Our workers don't have a voice in operational learning – we say they have a voice and often include them in our action items – but we usually bias operation learning towards planning and supervision.

By not creating a space for operational learning in our organizations and not giving the profound users of our processes a voice in operational feedback, we have actually reduced the amount of learning that we do. By moving problem identification upward, we have actually reduced the amount of information that we gather about our systems and process. Because we know less about our organization and its work practices, we, therefore, have less information to give to our decision makers. We make critical decisions with limited amounts of information. Knowing less does not make you smarter. Knowing more makes us all wiser. It is time to become wiser.

Leaders want to make the best decisions possible. In many ways, we are the data-collection providers for leadership. We owe our leadership teams the best information we can give them. If we give leaders better information, leaders ask better questions. If leaders ask better questions, they get better answers. Our job is to ensure that we give our organizational leadership the right information so that they then ask the right operational questions.

This book has one goal above all other objectives: we must give leadership the best and most complete information about our systems and processes in order to create an environment where our leaders will make the best decisions about our organization's future and keep our workers safe and alive.

I don't think there is much value in selling you on the idea that if you gather better operational intelligence you will make better operational decisions. I know you know this; at least I hope very much that you know this. I think what I can do is help you do better operational learning. I do this by showing you what we have learned about doing better operational learning. We know that learning matters. We know that management's reaction to learning opportunities matters. It would make sense that how you actually do operational learning activity would also make a difference.

## Wise managers make better decisions

In organizations just like yours, over time our leadership has developed a keen eye at identifying problems and creating solutions. Let's call this the find and fix mindset. Our organizations are filled with people who pride themselves at their ability to fix problems. Fixing is what leaders do. Fixing is how leaders become leaders. When we say "the buck stops here," we are saying that we are accountable and responsible for fixing problems that rise to our attention. If we don't fix problems, the problems most likely will not get fixed. If we don't know what the problems are, we are just fooling ourselves into believing that we are smart.

Fixing isn't wrong or bad. In fact, fixing is a good thing. Fixing is the point. The difficulties arise when we are fixing the wrong things. I would characterize the number one problem I see in organizations is that we fix things aggressively and completely when we think the fixes matter; the only problem is that we often fix the wrong things. We do it repeatedly. You have seen it your entire career. You have probably made jokes about this exact observation. You know as well as I do, you are part of an organization that fixes problems that can be fixed and often does not fix the problems that cannot be fixed, or it would be too hard to fix. In my history, I often observed that when one particular problem needs some attention, we usually try to fix all the people instead of the problem. If one worker sneezed – everybody was issued a tissue and required to take sneeze abatement training.

Our problem is not a lack of will or ability to solve problems. Generally, we love to fix problems. We are motivated to make the world more stable and better. We don't need to spend a lot of time asking our leaders to lead. We need to help our leadership understand not what they need to be fixing, but how they need to fix problems differently. We need to ask them to evolve.

Improvement is always a function of learning. Learning makes you more prepared to understand and facilitate decision-making. Learning makes you smarter. Knowing less does not make you more intelligent. Learning creates wisdom. Wisdom produces stable and reliable systems.

## *See hazard identification as an outcome*

We make almost exactly the same mistakes with hazard identification. We tend to focus on hazard analysis as a task to be done before work happens. We see hazard identification and analysis as an outcome, a thing that must be done. Sadly, that is wrong and it is wrong based on a flawed premise in our thinking and training. We see hazards as permanent and we assess risk based on that permanence. The problem is that hazards are not permanent. Hazards move in and out of work all the time.

Workers discover hazards as they do their jobs. We know that is true because we have so much experience recognizing hazards in our daily operational lives. If hazards where fixed, our analysis would be fantastic and we would have much fewer, if any, accidents in our operations. Hazard identification in a perfect world means find it, remove it, mitigate it, defend it, paint it, train on it, make a poster about it, write a procedure around it, and watch the safe and stable work start streaming.

The idea that permanent and fixed hazards exist in our production areas is way too simple. Hazards do exist in the production area. Hazards are everywhere, but the hazards are permanent in retrospect and discoverable when you do work. Take driving, for example, this is a task that probably most of us do. Hazards in driving are not permanent to the driver – the vehicle is moving – hazards appear and disappear all the time. The giant pothole on Main Street is permanent to the road, to the road crew, and to the taxpayers who fund the road. The same giant pothole is not permanent to the many drivers that must deal with this hazard in real time while operating a motor vehicle. Even potholes appear and disappear as we move up and down the road.

We can pretend that we will someday identify and analyze all the hazards that exist in our worlds, but that idea is not realistically possible and just wrong. We never stop identifying dangers and therefore we will never stop analyzing hazards. Hazards in our workplaces are fluid; they move in and out of the work we do. Identifying hazards is a process to be done, not a task to be completed. I know it makes work harder and less efficient for us, but I also know it is true. Risk-based approaches to hazardous operations are not about management of hazard; these methods were developed and used because they save time and reinforce the old idea that we can someday get them all.

Processes matter more than behavior matters. Process matters more than fixed hazard analysis matters. You don't manage outcomes ... outcomes are outcomes – the simple result of work being done – you manage processes. The process creates outcome; outcomes are nothing more than the logical output of a process. Outcomes don't happen in a vacuum. Outcomes are the products of structures, systems, and processes. If you doubt this idea, walk down the hall of your building and find the software programmers who work in your organization. You will hear these programmers say something about garbage going in leading to the garbage coming out.

Over the years, safety has firmly been based upon the idea that behaviors matter more than process. Couple this idea with a belief that hazards are permanent and knowable. We realize quickly that means matter more than behaviors.

These ideas are pretty controversial, and I don't understand why I figured out early as an employee of an organization that I was but a small part of a much bigger system. I think most people understand that. Certainly there are many studies to support that fact. What matters in the case of this discussion is the belief that by managing worker behavior (which I might add we are incredibly bad at doing), we are managing safety processes and practices. That idea is merely old fashioned and wrong. The idea of the "behavior-control-leading-to-better-practice" relationship does not exist and probably never existed. We are reasonably good at managing the process and stunningly bad at managing behavior.

What we need to do is to figure out how to think about and manage our organizations so that we don't use outcome information to solely formulate our ideas and methods we use in the management of safety and reliability. We have to build an understanding that if we look at outcomes as prediction tools for safety and reliability we are a day late and a dollar short. We must teach ourselves to study the process of safety – the practice of safety – not the safety outcomes that have historically, and quite powerfully, driven our thinking about safety. Safety is only one part of managing safety.

This historical bias says "that outcomes are what we manage" is even more dangerous when you think of the incentives that have arisen in the ranks of Senior Managers around safety as an outcome. Our organizations have permanently etched outcome metrics as scorecards for a manager's effectiveness. In a very real way, we can't be angry that managers are fixated by outcomes if we measure a manager's worth by the presence or absence of safety outcomes. The path to heaven is paved with good intentions or so they say, but are good intentions coupled with ineffective metrics any better than just good intentions?

# chapter six

# There is good news

> I never learn anything talking.
> I only learn things when I ask questions.
>
> **Lou Holtz**

These strongly held ideas that people carry forward for the old view of safety presents an opportunity to write a much different type of book. I address this question in the work I do every day. I talk to amazing people in amazing organizations that have the exact intention of doing the right things for their workers. These organizations have safety coded in their DNA. These are amazing, smart people who somehow do very counter-productive things in the name of making the workplace better – fantastic men and women who are leaders with the best intentions that make their organizations more dangerous, more silent, more opaque, and more uncertain in their attempt to create better safety outcomes. You must agree that something is wrong.

It is hard to let go of the old way of thinking because it was THE way of thinking. Getting your organization to think differently means that you have to first convince them to move away from their current way of perceiving the world and the workplace around them. This type of education is hard, but noble work and it is worth it. You will save lives. Don't stop and don't get angry.

How can we think about, talk about, and manage safe and reliable performance in a different way? Managing safety and reliability in a way that does not favor the outcomes of safety but supports the practice of safety in the work our workers do, both in failure and in successful work? How can we get leaders to think about safety differently, not as an outcome, but more of a practice? This urgent question has challenged me for a long time.

Let's set the stage with three basic ideas that I have found to be stable and reliable in my research of organizations that have sought to improve their safety and reliability performance by getting better at understanding safety differently.

Here is what we know:

1. All accident prevention happens as a result of learning what happens on the production floor, where the worker does the work. We must learn how work actually happens both in successful production and failure. Understanding work from the worker's perception changes the way managers and leaders think about work and work management.
2. Leaders, planners, and managers don't have the same perspective as workers when it comes to doing work – and unless something changes they never will. Why is this idea shocking? Because this point of view is incredibly different and that difference is very normal. The perspective we have when we observe our worker's performance colors the ways we perceive, think about and describe the work that the worker is doing. We bias our understanding, quite unintentionally, by looking at work from a perspective other than that of the worker. Planners see work differently than workers. Managers see work differently than workers.
3. Our investigations often wait for failure to create corrective actions based almost entirely on what did not happen – completely avoiding what actually happened. That is not a function of bad intent, quite the opposite; it is a function of perception bias. We naturally see the world, as we understand the world, not through the eyes of how others perceive the world. Same view, but a different vision.

One way we can shift from outcome to process and reduce the perception bias that we carry out to the production floor is to get better at recognizing that one group of employees, let's call them managers, are not any smarter (in fact might be less smart) then another group of employees, let's call them workers. In other words, our organizations tend to exhibit the belief in their management systems that managers are smarter than workers. Both of these groups of managers and workers have a way they see the organization and a way they are expected to interact with the organization, and one way is clearly not better than the other. It's not that one category of employee is any smarter than any other category of employee; the difference is in the perspective these employees bring to the problem.

I don't think we are wise enough to solve problems from only one worldview. It worries me that we don't include other perspectives and that we don't include those perspectives more out of our need for speed and efficiency – than any other reason. I am concerned that we don't widen our understanding to include processes and systems because we think that by broadening our understanding it will be too much. Too much work, too much information, too much time, too much disruption, too much. Sometimes the smartest and fastest guy in the room is not the wisest.

What we are doing is limiting our perspective while we are trying to understand our processes. I learned an essential and critical lesson early in my life; knowing more does not make you smarter. This entire discussion about outcome and process, the perspective of operations, and limitations of understanding all move me to make the point of this book. We have to be better at learning about our processes from the perspective of the people who do the work.

We must realize that when something bad happens in our organization, it is not a signal that our workers are suddenly doing badly. If a worker was trusted with strategic business information, given a set of keys to your building, or allowed to operate a multi-million dollar machine on Tuesday, what are the chances that his or her entire personality and motivation would have changed on a Wednesday morning at 9:52 am when a forklift was crashed into a wall? People struggle with losing weight, changing toothpastes … how could a person possibly transform from a skilled, experienced, and trusted worker to an evil, dangerous, and careless worker overnight?

# *chapter seven*

# *Why learning has NOT been our first, best tool*

How do you know what you don't know?

> When you think you know the answer there is no need to ask the question. The answer is the enemy of the question. This force of knowing is incredibly strong. Curious leaders are always much more prepared for both success and failure. Curious leaders are always learning and most leaders want to be curious.
>
> Here's the problem:
>
> The pressure to fix is stronger than the pressure to learn.

## *Let me make you breakfast: An illustrated discussion*

Let's imagine that I am going to cook you breakfast. Not just any breakfast, but the best breakfast you have had in a long time, the full western-type breakfast, with eggs, toast, bacon, oatmeal, fruit, yogurt, and a vast array of jams, jellies, and marmalades. It is a full-on breakfast extravaganza, and it is going to be good, in fact incredible.

You sit down and I start all the preparations. It seems a hundred activities are going all at the same time: cooking, chopping, simmering, boiling, toasting, and frying. Everything looks and smells great. You are starting to get hungry. The table is set and everything is ready. You are waiting for the food to be cooked and served. Breakfast is seemingly only minutes away and you cannot wait.

We are talking and laughing. You are reading the paper and commenting on all the news of the world. You have strong opinions about the world and this day's paper is giving you many topics around which you can expand in vivid detail. The conversation ebbs and flows, becomes interesting and animated and a bit heated at times. Breakfast smells great and can't come soon enough. So far, it has been a great morning with the promise of an even better morning when the food is eventually served.

Then you smell the unmistakable smell of burning eggs. No check that last comment; it is not the smell of burning eggs. It is the smell of very terribly burnt eggs. The eggs are ruined. The eggs are beyond recovery. The star of the breakfast has been disqualified from the starring role of the meal.

I am furious and upset. You are trying to both understand what has happened and de-escalate the consequences of the burnt eggs. Tensions begin to rise and disappointment fills the room. The breakfast that promised to be so good is ruined. The whole day has taken a turn for the worse. I failed at making breakfast.

We can now do a couple of things: We can go out to a local eatery and have an excellent breakfast out. We could put together an investigation team and try to learn what went wrong to ensure that this set of circumstances never happens again. One choice satisfies our hunger for bacon, the other satisfies our need to learn, understand, and explain how breakfast was ruined.

What is important is that this choice will either make us immediately happy or, after a brief period of learning, much smarter for the long game. One option gives us an immediate pay-off. Going out to eat is quick, satisfying, and puts the failure behind us. Staying and understanding how the eggs were burned is neither interesting nor fun, but it will eventually have some value – if we ever have breakfast together again.

This is exactly what operational learning boils down to in reality; immediate satisfaction and putting the problem behind us versus learning and improving.

The long game pays out well – just not immediately.

You can look at the infinite number of ways that breakfast could go right, but didn't – or you can see the actual way that breakfast failed. The importance is that these two choices are incredibly different. When you analyze what failed to go right, you miss what actually went wrong. What matters is what went wrong.

Learning is hard work and often not the easiest choice to choose to move the organization forward. It is much more satisfying to put the problem behind us, ignore what happened and go for the quickest answer. Knowing less does not make you smarter or safer.

All of your organizational events are information rich. All of them have information about your organization's systems and processes.

Every event has a story that it tells the organization, a detailed story that is filled with conditions that must have been present for the failure or near-failure to happen. Every event has a complete, complicated story leading up to the consequence of the event. Events are rich in context. How much we learn about that story is directly related to how we, as an organization, think about and respond to the event.

The information is there. The information is always there. Every event is information rich. All we have to do is look deeper to understand that context. We don't really want a *cause* for an event; what we want is an explanation of the framework of an event.

So, what is an event? The definition of an event is the degree to which an event is understood in retrospect. We tend to become very fixated on the retrospective understanding of something after it happens. The reason we are fixated is on the surface. After an event has happened, all we have is the retrospect of the event. Discovering and learning the context is not given to us as a present; we must go out and learn that information. Learning is work. That makes sense, of course, but it is also a bit late to get all that fixated. A bad thing, an event, has already happened. We really should have been exploring what COULD happen before it happened.

Outcome bias, the degree to which you allow your understanding of an event to be thoroughly informed of what has already occurred, is an extremely strong force in the identification of the problem. Because much of our analysis of an event happens post-event (virtually all accident investigations occur after the accident took place), our need to learn more from other people is actually quite low. I don't need to ask a lot of questions about problem discovery if I already know what the problem is. We identify what has failed after if has already failed. When you think about it, discovering some problem after the problem has happened is not terribly skillful or valuable in your operations, time or resources. It is oblivious, in retrospect, what the problem is afterward.

We are biased and we know it. We limit our need to learn about a problem by thinking that we already know what the problem is and we know that as well. Is all lost? Are we trapped in the human condition? How can we know we are limiting learning if limiting learning is a natural consequence of problem-solving? The answer to these questions is yes, but not a firm yes, more of a knowing yes.

The challenge is to create an environment for event communication that is fixated on operational learning and then allow this learning communication to happen. That open and non-personal learning environment is essential to understanding the story of how what happened, happened. This conversation is about learning the problem, not about fixing the problem.

Fixing is a retrospective response to an accident in retrospect. Fixing has to happen after learning, as a result of learning. Fixing can only occur after learning. Fixing that takes place before learning is a not fixing. When you fix before you learn you are not fixing what happened, you are fixing what you think happened in retrospect.

We don't learn very well. Alternatively, perhaps it would be better to say that learning has not been our first tool in understanding safety and reliability events. Learning is not valued as highly as fixing is valued. Either way, we are now beginning to understand that to move operational safety to a new and different level, we have to do some of our work differently. We have to change the inputs into our safety systems in order to change the outputs of our safety systems.

To have this discussion, we must first go back to some basic ideas in the new view of safe and reliable operations. We must first try to understand what makes up an event. Each process, system, production, or even personal failure is an event. All of the things that happen in your organization are events with or without consequence. Events to some forward thinking organizations may not even have failed yet to be an event. Prediction, near miss reporting, and low-level problem identification usually identifies some type of an event.

So when is an event and event? Well, the quick answer is that we must look beyond the usual subjects. We know that the operational definition of an event is actually a failure viewed after that particular failure has already happened, a failure in retrospect. I am not sure that the definition of an event has any value to our organization operationally or warrants any special treatment.

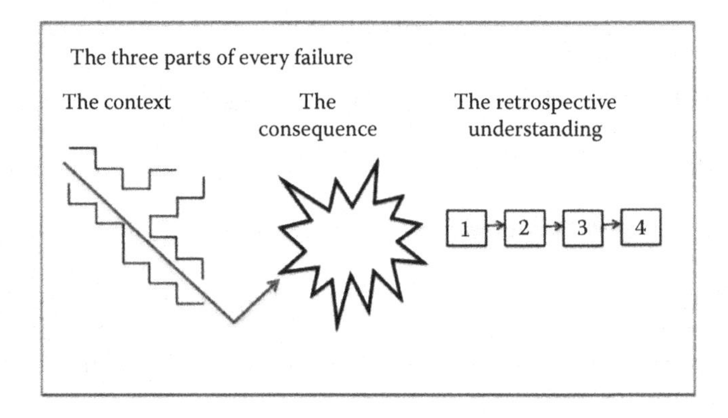


To better understand this let's discuss events by taking the concept of an event and breaking it up into its parts. An event is not one thing, an event is a combination of several things that could happen, have happened, and the way we think about what happened after it happened (or nearly happened).

All events have three distinct parts: 1. The context, that is everything that took place before the event happened. 2. The Consequence is what happened, nothing more and nothing less. 3. The Retrospective way the organization views the event, post-consequence, that is how the organization simplifies and linearizes the event into an understandable and analyzable problem to be fixed.

Our challenge is that what we most likely believe is the most important part of the event, the consequence, is in reality the least interesting part of the event. We treat consequence as if it were some interesting and significant piece of information. I understand why we do this; the consequence is often quite terrible and is always operationally dangerous. Just because the consequence is provocative doesn't make that specific outcome more or less interesting or informative.

I would suggest that the consequence is not exactly that important for three reasons.

1. Reason One: The result has already happened and, therefore, there is little we can do to make the consequence any different. We don't have the power to turn back time or change history. We cannot go back and undo what has already happened. In a way, the consequence is on the wrong side of the equal sign for us to have any power over what it means. The best we can do is managing the message around the consequence. Often times the way we handle that message is to determine blame. I have found that usually the last person to touch the "thing" is usually determined to be the person who we blame.
2. Reason Two: Consequence is the only part of a human to systems failure that actually obeys the laws of physics. The consequence is seductive to us as investigators, managers, even learners because we get to use our engineering worldview to talk about exactly what happened in a very mysterious and confusing human-centric world. We can deal with facts – hard, cold, facts. We don't have to deal with the fluffy stuff, the non-fact based parts of an event. The notion that the consequence is explainable using technical terms is very attractive. It is also very uninformative and potentially misdirects the learner towards a very simple understanding of the workplace world. If you drop 500 pounds of steel on a worker's foot, you will smash the worker's foot. If you wanted to invent a foot smashing machine what would you need? 500 pounds of steel and a worker's foot, not very interesting is it?
3. Reason Three: All events are information rich. Sometimes a big event gives us many individual small learning opportunities. Sometimes minuscule events can give us significant information. Simply dictating the level of learning based solely on the consequence seems a little too easy. We must be curious enough to know that we don't know what we can learn until after we start learning. Casually choosing to not learn because the event does not meet some harm threshold seems a bit wed to some type of artificial standard.

## *Not every event needs fixing*

Some big events have limited learning. Some small events are so information rich that the learning seemingly never stops. Knowing which event will give your organization real and tangible information is difficult. Coupled with the problem of not knowing which events have learning potential is the absolute fact that some events simply are not incredibly interesting. We need to be able to tell the difference between the two choices.

The error that has little to no consequence does not require much of your attention. Even though we know that little failures can serve as small signals to larger failures, not every little failure is critical. Knowing when to call in the resources is much more of a technique. We are always doing small learning. Curiosity is a tool used wisely and appropriately.

One strong delineator between these two modes could be the amount of value you believe the event may have for the organization. If you as a leader find the event interesting, it is probably kind of interesting and warrants some learning time. If your workers find it interesting, that is even more of a reason to begin learning.

To be fair, not every event needs a full investigation. Your organization would not be able to afford that level of effort. But assuming that decision is made better, or worse yet easier, by some artificial standard of importance based upon consequence is a simple way to simple.

Be curious about how the event happened. Trust your instinct and ensure your organization can explain the event to better learn from the event. Our goal is to learn so we can manage the conditions of the event out of the system.

We must learn to live on the left side of this model ... when you learn to live on the preventative side of the consequence, the world looks different. You change the way you see the world and therefore you change the things you learn. Our goal is to focus learning on the context before the consequence – the context is the only place you can have an impact on improving your organization. Knowing how the event happens is much more important then knowing what happened – because you already know what happened – it happened.

Learning lives on the front side of the consequence, not on the backside. The problem is our tools have forced us to spend the majority of our time focused on the actions through the lens of the consequence.

When we say, "A bad thing happened because the worker did not follow a rule," we must also ask this second question, "When good things happen does the worker follow the same rule?" Chances are statistically high (like 100%) that when success happens the worker is not following the rule either. That makes a compelling case that the problem is not rule-following (post consequence thinking), the problem is the rule or system (pre-consequence thinking).

That is a much different conversation. Our entire goal for improving operational learning is to ask different questions. The power we have as guides and facilitators of safe work capacity is not in the answer to the question. The power is in the question.

# chapter eight

# A learning team case study

**Bob Edwards**

> The case studies used in this book are all provided by Bob Edwards. Bob is an early leader in using Learning Teams in lieu of typical investigations. I asked Bob to provide some case studies for us to use to help illustrate the power of operational learning in a real workplace.

## *Free Willie: A case study for learning teams*

A good person who made a bad choice is no way to think about a worker who has been involved in an operational failure. Saying a good person made a bad choice is just another way to say "bad person."

Willie had been a fork truck driver for more than 20 years. He was a hard working employee and came to work every day on time. His job was an incredibly busy job and he did it well. He maneuvered his fork truck with skill that only comes from years of experience and skill. It looked more like a choreographed dance routine then it did work. There is probably not enough paper available to proceduralize the art of doing what Willie did.

His primary responsibility was to load product all day into the back of semi-trailers. Sounds simple enough unless you had the opportunity of trying it sometime. He had to stack the product in a specific way. Each product group had to be stacked in its own organized nuances and requirements to be successfully transported. Each product was serialized with a barcode to make it possible to track it from the production floor to the customer's home. There were numerous loading dock doors that Willie had to load into and placing the correct product into the right trailer was imperative. As each pre-stacked group of products was loaded into the respective trailer, Willie scanned the information into the wireless computer system to ensure the work was completed as planned. Many of the pre-stacked products were such that as he loaded them into the trailer, there was only an inch or two clearance on each side between the stack and the inside walls of the trailer. For anyone who has ever seen this type of work performed, you will know that visibility and lighting are also a big part of the challenge.

As Willie completed the loading of each trailer, he reached a point where there was just enough room for one more stack of product to be loaded on the back of the trailer. It was important to fill this last gap with the product so that the trailer could carry the full capacity of products and maximize the shipment. Also, if he did not place the last stack in, he would have to install crossbars to keep the product from falling over during transport. Therefore, it totally made sense for Willie to fill the trailer all the way to the back edge, and Willie was an artist at doing just that. Well, it turns out that there was a slight problem with loading all the way to the back of the trailer. A problem that Willie and the team had overcome. You see, if they loaded all the way to the back, the last stack of product would set on top of the edge of the dock plate. This made it almost impossible to get the dock plate up when it was time to release the trailer for shipping. In case you don't know much about loading docks and trailers, I will give you a quick lesson. When a semi-truck backs a trailer up to a dock, several things have to occur before loading or unloading the trailer. First, the truck driver has to open the doors of the trailer and carefully line up and back the trailer up to the dock. Once he has the trailer in place, a locking device is activated that grabs hold of the metal bar that is located at the back of the trailer directly below the back door opening. This locking mechanism prevents the driver from inadvertently pulling out before the load is ready to be released. Next, the dock personnel will install the dock plate. This device is a large metal plate that comes out and covers the gap between the dock and the trailer so that the fork truck can successfully drive into the trailer to either load or unload the product. In addition to the dock lock and dock plate system, there is a light signaling system on both the inside and outside of the dock. The light on the outside shows green when the dock is ready for the truck driver to back the trailer up. This indicates that the dock lock is turned off and the dock plate is out of the way. Once the trailer is in place and the dock lock and dock plates are in place, the outside light turns red indicating that the trailer cannot be moved. On the inside of the dock, the opposite situation exists. When the light on the inside is red, that indicates that the dock lock and dock plate are not engaged and it is not safe to enter the trailer. Once the dock lock and plate are in place, the inside light will turn green indicating it is safe to enter the trailer. It is crucial to follow the signaling system because there are potentially severe consequences if a fork truck driver is driving in or out of the trailer and the truck moves out. At this particular site, there was one small issue with the light signaling system. When the dock lock was engaged, the inside red light would turn off. However, the light did not turn green until the dock plate was in place. This intermediate step left the light completely off until both the lock and plate were in place. The fork truck drivers were to assume that "no light" was the same as "red light" and they were not to enter the trailer until the inside light was green.

So now, back to the story. As I mentioned earlier, Willie and his team had come up with a solution to the dock plate issue. Instead of setting the last row of products on top of the dock plate, Willie lowered the dock plate out of the way and then would drive up to the trailer. He did this without driving into it and set the stack right at the end of the trailer. This made it easy to finish the load, close the trailer, and send it on its way. On one occasion, Willie had just completed loading a trailer using the procedure described above and had moved on to work on the next one when he received a call over the radio. There was a problem with one of the scanned products from the previous load and the dispatcher asked if Willie would please unload the product until they reached the missed-scan unit. No problem, Willie simply picked up the last stack that he had put on, and set it to the side of the dock opening. Then something tragic occurred. Willie returned to pick up the next row forgetting that he had not re-installed the dock plate. He drove across the opening into the trailer and immediately fell into the crack between the dock plate and the trailer.

Now before we go any further, it's important for you to know that no one was injured and no equipment was damaged. However, if you know anything at all about fork trucks, you will know that they are not all-terrain vehicles. So now, Willie was stuck and so he signaled for help. It was not a major deal to remove his fork truck; his co-workers simply attached a towing strap and used another fork truck to give him a little pull to get him back out of the crack. Now they had a bigger problem to deal with. You see the site had some strict rules on fork truck safety. One of those rules was that you never ever enter a trailer pulled up to the dock unless the dock lock is in place, the dock plate is installed and the light on the inside is green. So from the leadership's point of view, this was a clear violation of company policy, and, therefore, a violation worthy of punishment. With no real learning attempt at all, the decision was made to administer discipline. Willie was written up with a final written warning, sent home for three days without pay, and his safety performance award was taken away. And just like that, Willie went from a valued, long-term problem-solving worker to a villain on the verge of losing his job. Now stay with me and don't get too mad. The story has a happy ending. This event occurred early in the site's journey on improving their understanding of human performance and the need for operational learning around failure, so their reaction was not really a surprise. When Todd saw this event, he realized that this was a golden learning situation to help the leadership team begin thinking and responding differently to failure.

Todd was at the site training the leadership team on how to improve their response to failure when this event occurred. What an excellent coincidence. The safety team brought the event to Todd and explained what happened and the actions that were taken. Todd's response caught them off guard. After listening to the story, he got up, closed the door, and began to talk about how the leadership appeared to be punishing Willie for what was actually human error. He told them that punishing Willie for doing something that he didn't intend to do in order to make him come to work tomorrow, and not do what he didn't mean to do today was not fixing anything. The safety leader tried to explain to Todd that the safety rules around trailers and fork trucks were there for crucial reasons. He talked about being firm, fair and consistent and about getting the fork truck driver's attention about how serious this could have been. Todd didn't waiver. He didn't change his stance. He just kept bringing the leaders back around to the fact that punishment should only be used for times when employees deliberately break company rules or policies. It should never be used for human error triggered events. Todd finally asked the leadership team if they would be willing to try out a Learning Team approach to this event. They agreed.

First things first, they had to bring Willie back in off of his disciplinary leave. Todd made sure the leadership team understood that you couldn't do a proper investigation without the person or persons closest to the event. You simply will not get the holistic picture and deeper story unless you include them. The goal of bringing Willie back in was not to interrogate him. It was to have him explain how he did his work and for Willie to walk the team through the conditions and path that led to the event. So Todd pulled together a team to learn. Oddly enough, this was the site's first official "Learning Team." The goal of the Learning Team was to take some time to have an open discussion on the conditions and causes that led to Willie's event. The Team included a couple of Willie's fellow drivers, his front line supervisor, the site safety guy and of course Willie. They set up a flip chart in a small conference room and Todd facilitated the conversation. The safety leader was expecting a traditional "interrogation" styled investigation using something similar to the 5-whys or some sort of reward look at the failure – some means to quickly get to the root cause and to justify the disciplinary actions taken. Little did he know that he was about to experience a transformation in his thinking approach around failure. Little did he know that he was about to embark on a new path in his career?

The team was pulled together, except for Willie. He was a few minutes late understandably so. He was just suspended for his "bad" actions and was probably a little concerned that he was about to get another dose. So when Willie finally walked into the room, the first words out of his mouth were "I promise you all this: I will never do that again!" Let me ask you a question. Do you believe that he meant that? I do. Do you think he can promise that? No way! How can he promise to not do something that he didn't mean to do in the first place? Todd handled it well. His comment back to Willie was, "Thanks for promising us that, but I don't think you can make that commitment. I don't believe you came to work intending to drive your truck onto the trailer without the dock plate in place." Then Todd began the discussion. He started by asking questions to Willie and his fellow drivers about how they load trucks ... the way they usually conduct the loading and the types of things that come up that cause them to have to adapt and adjust. As he continued to ask questions, he just wrote facts, conditions, issues and work methods down on the flip charts, and they began to build a sort of wall of discovery. The more Willie and his fellow workers talked, the more it became apparent to the whole team that this was a human error that triggered an event with many complicated pieces that led to the failure. The more the team looked at all of these conditions, the more they saw that punishing Willie was not going to fix the system weaknesses. The team was able to provide a better explanation to management on how the system weaknesses, when coupled with human error, led to Willie driving back onto the trailer without the dock plate in place. Therefore, the punishment didn't fix anything at all. In fact, the process was such that it would not be even a little bit surprising to see the same sort of event happen again in the future. Drivers still have to adapt when issues come up. The other drivers are still taking the dock plate down to load the last row of product. The light was still a small round light up on the corner area of the dock door opening. The light still changed from green to nothing in the state where the dock plate has been removed and the dock lock is still in place. And it is probably noteworthy that there is another red light on the dock that is used for a quality signaling device and is located right next to the dock signaling light, providing additional confusion. The fact is that most of the time the drivers just simply load the trucks and send them on their way. It was rare that they had to unload what they had just loaded and yet the possibility still existed. The actual potential for this mistake-triggered event to ever occur again was inevitable. Unless the leadership team takes some immediate action to actually change the way this work was being scheduled, managed, measured, and done in order to make the process of loading and unloading truck trailers more reliable and resilient, another accident would occur.

After a couple of the learning sessions, the team decided that they would recommend to management that this was not an intentional rule breaking situation, and, in fact, was a human error triggered event that could occur again if the systems were not improved. Now came the real test. When management was armed with the operational intelligence that showed them the deeper story, the story of complex adaptive behavior in an extremely flawed system, they were faced with some serious decisions. Should they admit that they reacted too quickly? Would they admit that they should not have administered discipline? If they admitted that they were wrong, should they change their ways moving forward, or go back and change Willie's record and remove the punishment? Well, they needed a little "soak time" to think it over, and so Todd didn't push them to make a decision right then. He suggested they regroup and discuss the following day. What they did next was quite remarkable and changed the course of actions for the leadership team moving forward. The management team decided that if they were going to learn to respond differently to failure, now was as good of time as any. They chose to give Willie his money for his lost days while out on disciplinary leave. They gave him his safety award back. They removed the discipline from his record, and they published the story in the weekly newsletter that goes out to the entire plant. They didn't really have to publish it in the newsletter or on the website. Word of mouth (the rumor mill) had carried the story to the far corners of the plant already. The management team was changing how they responded to events. They were listening to the entire story; they were talking more openly with the employees and, in fact, were determined to stop emotionally reacting when something bad happened and instead take the time to learn and respond. The site was now officially on their way to better operational learning and a more holistic, non-blaming approach to responding to failures.

An interesting thing happened next with this Learning Team. The same team that told the story of "how" the incident occurred added a few team members and began to discuss what would make the system safer and less error prone. As they discussed different ideas, it became evident that they were coming up with ideas that would actually make the loading process a lot better. They decided they needed to fully interlock the light system so that there was never a condition where the lights went off entirely. If either the dock plate or dock lock were not engaged, the light would turn red. The team also decided that the small round light at the corner of the door opening was not enough. They ordered, and had installed a large L-shaped LED light bar. It was installed at both top corners of the dock door opening making it very easy to tell if it was safe for the driver to enter the trailer. They even set the light system up so that when it was in the "red" condition it flashed on and off continually and when it was in the "green" condition, it stayed on constant. This feature aids with any drivers who may struggle with color blindness since the site does not discriminate against that condition. The team also decided that they needed to use a new style of dock plate that has a small "kick" plate that prevents drive-offs. This allowed the driver to lower the dock plate to load the last units on the trailer but prevented him from being stuck in the gap between the dock plate and trailer. The team also decided that it didn't make sense to send Willie back to "How to drive a fork truck" training. Instead, they considered his time helping with the Learning Team as the fulfillment of the "re-training" requirement. At the end of the day, the loading system is much safer and error-tolerant than ever before. Next, the site began its task of modifying all of the dozens of dock doors around the facility. Although that took time, as they made the modifications, they were in fact making the system much better than if they had made their focus on merely counseling, retraining and disciplining Willie. Now Willie has become part of the solution set to a previously weak and error-prone system. And best of all, Willie was "free" to return to his work to load more product in the back of trailers. All day long!

# *chapter nine*

# *Why we don't learn?*

> Even though there are no ways of knowing for sure,
> There are ways of knowing for pretty sure.
>
> **Lemony Snicket**

There are many reasons why we aren't good at learning about our organizations.

We are often in a hurry to answer the immediate, burning question. We always have the need to make our systems stable again as fast as possible. We often have people that we must notify when something happens, usually people that are important in our organizations. The pressure to fix what happened overtakes the time-consuming process of knowing what happened. But, mostly, we stop learning when we think we have learned all there is to know. We don't need to learn what we already know. We already know what happened.

In many ways, we are not very modest, not very curious. We have a history of being arrogant. What interests me is that I can't figure out why we don't ask more questions. It almost seems human nature to ask questions. It appears, at least on the surface, that one of the reasons we are not more curious is that we tend to believe that we already know the answer to all of our questions. That operational immodesty, the lack of curiosity, has very negative consequences for our corrective actions, our operations, and the way our organizations function. If we are arrogant enough to believe that we already know the answers, there is no need to ever ask any questions. We don't learn because we don't think that we need to learn. If you already know the answers, you don't need to ask any questions.

Organizations are biased to believe that the worker is the problem. For some reason, a good worker became operationally incompetent. That bias is incredibly dangerous for three reasons: 1. It is wrong almost all the time. 2. It causes a wrong type of thinking about the problem. 3. It doesn't solve the problem. When we are convinced that the problem is the worker's behavior, the learning tends to go out and seek places where the worker's behavior should have been different. Those places are very easy to find. When you find these places, they are attractive to the investigation team. All that good thinking about doing a great learning is misdirected towards a temporary and non-sustainable understanding of what should be done to fix this particular worker in this specific situation. This investigative bias towards correcting the person actually tells a different story of the event. Usually, it is an incomplete story. It provides a false sense of resolutions as whatever caused the problem.

Think of the safety journey as one of the vast improvements. When we first started managing safety, we did it by making and enforcing rules about safety and we got better, much better. Then we started to level out; we reached a plateau in our performance and needed to do something different. Safety got better and then it stopped improving. We were better but not excellent. And, we did not solve all of the underlying unidentified problems. The next step was our beginning understanding of safety by design; process safety ideas started to break our organizations out of the performance stagnation, and we got even better and safer. We could develop systems to not hurt workers. The design was an incredible tool and we got even better. We now are at a place on this safety journey where more rules and more design do not equal more safety.

Organizations tend to improve quickly at first because there are many little things that can be fixed, monitored, or improved with little-to-no effort. These "high number of event and quickly improving companies" (the get-it-done crowd) tend to reinforce safety compliance, heighten workers attention with either strong positive or negative incentives, and increase management attention to safety.

## *Workers must be involved in problem identification*

> If you want to know how work is done, who should you ask?
> The answer is the worker.

## *Learn and improve*

Recently I have begun asking managers why their organizations do investigations. It is not a difficult question; at least I had never thought of this question as being the type of question that could stump a room full of high-level managers. I was simply wrong. I have had numerous instances where, upon asking why their organization does investigations, I was not given a quick answer. Sometimes leaders will say that the goal of an investigation is to stop repeat occurrences of an event. Many times the leaders just look at me and say nothing. This isn't a difficult question. The only answer to the question of why your organization does investigations is to learn and improve.

## *Workers are fundamental*

Problems are difficult. It means that something is not working the way you imagined it would work. I am confident that most managers don't want problems in their operations but are mature enough to know that problems will exist. In fact, many managers fancy themselves as excellent problem solvers. And, in fact, most managers are good at problem-solving. They have all the resources; therefore, they have the essential ingredients towards getting solutions. There is a real relationship between problem-solving and organizational power. I solve problems; therefore, I am both powerful and vital. I guess that is true, in a way; however, it is just an incredibly egotistical way to view operational leadership.

In reality, workers have so much more information and hardly any power. Workers must be included in investigations and learning activities. Without the worker as a part of the Learning Team, you are actually making it more difficult for your organization to fully understand how an event happened. If you do nothing else as an organization, start including the workers involved in the event (or a near event) in the investigation and learning about the event.

The competent workers have a vested interest in ensuring that your organization learns how events transpire. For some reason, fear must be that the affected workers may not tell the truth, that they will somehow be driven to hide the real story or tell lies about the event. That may happen, and if it does that is much more of a problem of your organization's culture and ability to learn then it is a worker trying to commit a crime and lie to you. I really don't understand where the need to separate the worker involved from the telling of the story of the event began, but I know that we must work hard to stop this practice.

Given the fact that your workers are not criminals or sociopaths, these workers will have all the details that happened during the event. They know how the event happened; therefore, they are vital in the collection of information, telling the story, understanding the event, and helping you fix the problem.

The problem with learning about problems is not in the lack of desire to deal with problems, but in the almost extreme pressure to correct the issues. We live in organizations that value outcomes (problem solutions), not processes (problem learning). We see our jobs as "fixers" of problems. We are rewarded and recognized for having solutions to big, hairy problems. Because the answer is so important, we tend to rush into solution mode. You see that in our formal problem-solving processes and you hear it in the language we use to describe problems and solutions. In reality, what a manager has been told through years and years of leadership training and leadership culture is that good managers actually only need to know three important things:

1. What happened?
2. Who did it?
3. How did you fix it?

Sadly, none of those three choices involves or encourages learning. They are stagnant. If the questions don't create a space to learn, information from the worker's perspective is non-existent. In fact, these three choices actually discourage communication and learning and emphasize fixing. This is a manager who doesn't want you to bring them a problem. This manager only wants solutions. This very rigid performance expectation, this rule for visiting his office, is frightening and wrong and it happens all the time.

I used to work with a senior leader who had a sign on the door of his office, a big fancy sign, which asked those three questions above and added a warning that the three questions and the three answers must fit on one power point slide. That manager was arrogantly proud of this expectation. It was a rule that must be followed. This manager was so convinced that this method was correct he would not meet with any of his leadership team until they could meet the three-question/one-page criteria. He often sent people away until they were able to come back with the single page overview of the operational problem. I am sure that he thought he was making the world a better place. He was cutting out the non-essential information and getting right to the root cause. He was the type of no-nonsense leader the world needs more of ... except the world does not need more of this offensive oversimplification of problem identification and complexity; it actually needs much less. This kind of manager is as dangerous as the problems that are happening on their watch.

What this manager was doing was forcing his organizational leaders into a position where they were completely incentivized to create a single, stunningly over-simplified answer that was immediately fixable, while affixing blame on some person (usually in this case the last person to touch the problem got associated with the blame). I can't think of a worse or more dangerous way to run an organization. He completely cut out any flexibility or latitude for deeper understanding of the story of the problem. He did it in a formal, almost gate-keeping way that actually restricted his people from bringing complex, multifaceted problems for further discussion and understanding. In fact, he had created an organization that devalued learning and emphasized speed and assumptions about processes. He truly was a manager that only cared about outcomes, and he limited learning to as close to zero as I have ever witnessed. He is the type of manager that should go the way of the dinosaurs and the Dodo.

## *Quick! Fix versus fix quickly!*

The pressure to fix outweighs the pressure to learn. That pressure is intense. It is so strong that our discussion of including workers in problem identification had to begin with a pre-discussion of the pressure management feels to provide immediate, if not very well thought out, solutions to all problems.

Managers know that workers have knowledge that they want. Managers simply don't know what they don't know. If a manager knew that a worker had knowledge of a process or system that was especially dangerous, it is a good bet that the manager would ask the worker to explain what is going on at the production level. The problem is, the problem always is, and we don't know what we don't know. If this fact were true, then one would think that managers would be the most curious people in our organizations. In reality, managers should be intensely interested. The problem is that if they believe the system is functioning well – they then assume there are no problems in the operations of their system. That notion that "no news is good news" is almost unanimously wrong.

Problems to be understood are always lurking and exist in all processes and systems; problems are not the exception to the operational rule, and problems are the rule. Problem identification and solution, real-time detection and correction, is consistently happening. In fact, detection and correction are never *not* happening. One of the challenges we all face organizationally, is that problems are so routine that sometimes substantially significant solutions to these problems, solutions that would typically trigger some type of change management threshold, are so normal and every day that the worker often does not even realize a significant change has been made.

There is hope, however, and hope's name is Bob or Jim or Janet – and they work for you. Your workers know your processes. In fact, they are experts at how you do your work. They are insiders, inside your system and processes, and inside your production areas. They know where your work systems are solid and recoverable and they know where your work systems are weak and brittle. Better still, they have known this information a long time. They use this information to give you the very outcomes you measure and desire. Nobody is smarter than your workers about your system.

We often discuss the model that illustrates the difference between work as imagined, planned, and work as it is actually and realistically accomplished. You know the difference between planned work and actual work exists because the difference between planned work and actual exists in your operational world as well. You have never done a job as planned in your entire career; every job contains a certain amount of variability and difference. It is the nature of adaptability. You may have your to-do list and a plan to attack it every morning, but events, impromptu meetings and phone calls in your day make it fluid. Your system is full of change and adaptation.

Workers that are involved in problem identification know more about the problem than almost anyone else in your organization does. These workers bring experience, understanding and expertise about the processes in which the work actually happens and knowing that information can usually improve. They enlighten the depth of problem knowledge, problem solution, and operational reliability. Knowing more is an influential position for you and your leadership to place yourselves. It is game changing. You want to know as much about the problem as you can know. You always want to know more as opposed to knowing less. Knowing less does not make you smarter.

Workers also tend to be much better in identifying the problem in the context of your work and your workplace. Workers tell the story of how work happens. That story, like all good stories, has a beginning, middle, and an end, moves logically through time, and contains nuances and complexities that only exist in the work itself. The story is how work actually happens. The story is everything about your operations. The story is crucial, foundational to your success.

The happy outcome of that experimentation is not only solutions to problems, but perhaps more importantly the identification of problems. You almost never solve the problem you set out to answer; you most certainly discover an incredibly different, much more foundational problem that needs to be resolved. That problem-switch is so reasonable that organizations that are good at learning expect the problem to change. In fact, they are often worried if the Learning Team does not come up with another problem area in their learning process. The problems you solve will not be the problems you set out to answer and that is reasonable and appropriate. And the fantastic news here: it will make your organization safer and you a lot smarter.

Workers must be involved in the problem identification activity. Workers should not just be looped into this discussion after the analysis has been done as some type of inclusion activity; this is too little too late. Workers are the vital, vocal majority in your operational learning process from the start to the finish.

In organizational learning, we have found the greatest learning happens when the ratio of workers to supervisors is about five to one or greater. Five workers to every one supervisor on a Learning Team seems to be the best way to give worker input the strongest, most active voice. In fact, many teams are quite functional without the presence of any supervisors.

The workers understand how work happens in practice and in reality. Your workers know which of your processes work and which of your processes are not working. The workers know how work happens in almost every possible condition, good, bad and ugly. It is the fluid work that takes place daily in all kinds of conditions. Your workers are, in many ways, brilliant at understanding and using your systems. In short, your workers are smarter then you are at understanding how your organization functions. You are fantastic at supervising, but they "get" the work.

## *By giving up control, you gain operational intelligence*

It is hard to symbolically hand over the reins to your workers when you have an operational problem that needs to be solved. It takes an incredibly trusting and mature manager to realize there is wisdom gained by not knowing the immediate answer. In fact, leaders all over the world are starting to realize that the real power a leader has been in not knowing the answer, but in understanding the right questions to ask. Leaders must trust that the workers, the people who own the work in practice, know more about how the work is done then the manager ever will know about the work.

## *Micro-experimentation*

First problem-learn, then problem-solve. Start by establishing what you don't know and work with that "not knowing the place" as a starting point. We must learn what the question is before we can answer the question. We must learn before we problem solve. Don't go into a problem trying to pre-determine the problem. I work with many people that will want to have every document from an event as a pre-read. It is a form of control – in order to better prepare themselves for the event review, these folk feel they need to have read and understand all the written information that existed before and after the event. In my opinion, pre-reading anything is not a good idea in that the pre-reading activity starts a whole series of early and strong biases of thinking. It is just not effective. In an actual real way, the learner is going into the learning event with a pre-conceived idea of what happened. The problem is that most of these preconceived ideas are usually wrong. Our job is not to put limits on our ability to learn by believing that we already know.

We structurally limit learning; learning leads to the opportunity to run small tests of new ideas for improvement. We learn, we try out our learning, we collect data, toss out what does not work, save and improve what does work, and we get better. I am certain you know this cycle and that this cycle helps in building our case for operational learning at the worker level. Organizations that are magnificent at learning are good and quite experienced at performing and monitoring small experiments. In these learning experiments, we quickly discovered that the workers are vital to understanding where the problems and conflicts exist in your organizational processes.

Workers are very aware of what parts of your processes are reliable and effective and which parts of your processes are especially dangerous. In these experiments, workers are required parts of the cycle. Doing a work activity, one time is very different than doing that same job activity hundreds of times a day. The workers run the tests and give the feedback. In doing so, the workers become smart and understanding not the answer, but in understanding the problem. That difference is not subtle and should not be lost. The prize is not in writing the perfect corrective action; the prize is in asking the perfect question.

## *Confidence is important*

It must be clear to you by now that the key to reliable performance is the engagement of the workers in problem learning (problem identification) and problem-solution (determination of corrective action).

The first and most important step to creating real worker engagement is building the worker's confidence in their belief that they can make a positive and lasting impact on your organization. The key is to build trust and confidence in your organization's ability to learn and improve for itself. No one knows better how your work is done than your people doing the work. This group must believe they can make a difference in the outcome of the learning activity to learn and own the corrective actions needed to make the process better.

Confidence is different from over-confidence. Think of this comparison: Being over-confident is exemplified by fixing the problem before you do the learning. Confidence is the ability to know that you don't know what has happened until you discover and understand.

## *Access reality when learning*

Think of your job like this: Your bosses make the best decisions when they have the best information. You make the best decisions when you have the best information. We owe our management the best operational intelligence that we can possibly provide to set up our managers and our organization to be successful. Good quality information as an input – equals – better corrective actions and process fixes. In many ways, you are giving a voice to the worker. The truth of your operations lies where the work happens, not in planning or managing work. Accessing that truth is critical to knowing how an event transpired.

## *Learning should be simple so that learning outcomes are not simplified*

Learning is messy. Learning can be complicated. People can get emotional, angry, upset and offended while learning is happening. I don't think I have ever done an investigation where we did not have some type of argument transpire during the work. Learning is hard to do because it is not quick, it is not clean, and if you are good, what you learn is often quite passionate. People can easily mistake event learning with the preparation steps for assigning blame and leveling accountability. I am convinced that some investigations do just that, move responsibility from the organization to the worker. Learning must be easy and accessible because the things you are going to learn are often complex, deep-rooted, and difficult to immediately identify.

## *Learning happens on a diffusion cycle*

Think of operational learning on an "S" curve or perhaps better still, a diffusion curve. The Diffusion curve, a concept developed by Everett Rodgers, talks about the types of people who accept innovation easily (early adopters) or the types of people who accepting innovation is much more uncomfortable (laggards). The same is true with operational learning. Some information is easy to identify early in the quest for an explanation. Some information in learning is much slower in appearing before the team. Just as with the Diffusion of innovation, there is a way information flows with time.

The fact that learning only gets better by investing more time for learning is critical to understanding the operational improvement. Knowing more makes you smarter, but you don't start out knowing more. You start out knowing little. You know the least about an event when the event happens. Every moment you spend learning how the event happened makes you and your organization smarter. When you are expected to know the most about an accident (the moment after the mishap has transpired) is the time you know the least about the accident. This is the irony of the learner and this irony has colored the way we learn and give answers.

This is important and a bit difficult to understand at first. In a way, this is like how you identify an expert in some work. To truly recognize that a person is an expert, you must first have some expertise yourself. To know that you don't know something takes the knowledge that there is more you need to learn. It is hard to not know what you don't know, and yet you still don't know what you don't know. This is all confusing, but still important. You must know that you don't know to begin allowing learning to happen in your organization. This becomes a bit of a "what is the sound of one hand clapping" discussion. Stay strong, the more we discuss this idea, the clearer this idea will become.

There is a definite beginning to learning. This beginning of operational learning is best characterized not as learning, but as a notification of an event. When an event happens, there absolutely is a need to tell people in your organization that something has happened. This notification process is vital; leadership should never be surprised by an event. The profound difference is that notification is the period where you, the notifier of the event to management, have the LEAST amount of information about this event. You know little and what you do know is most likely wrong.

For too long we have had the mistaken impression that when an event happens we must immediately know how to fix the problem. This is not only wrong thinking but also remarkably dangerous.

There is a "middle place" of learning – a place where learning happens and apparently gets to a point where there is not much more we can afford to learn. This middle space is where you gather information about how the event transpired and analyze this information to determine what your organization should do, how you should act. This middle area contains a period of time where you learn much detail and explanation and gather much information. This information will come from multiple places and provide substantial insight about the conditions present in the failure your organization has experienced. Soon you will learn much and the information will begin to taper off. The "s curve" will start to flatten out. You will have gathered sufficient amount of information to begin understanding and explaining the event.

This is the time to stop learning and start acting. This section is the report and a correct section of your action. You know what happened and can explain what happened. That explanation of the event is the report. That is your formal and documentable understanding of your event. You also are in a position to either recommend corrective actions or, better yet, begin your corrective actions.

All three of these parts of understanding and learning about an event are different. Perhaps most important is the idea that as you move through these three modes: notification, learning, and fixing, the knowledge you have about the event gets deeper and more explanatory. You are more intelligent because you know more, abundantly more, about what and how the event happened.

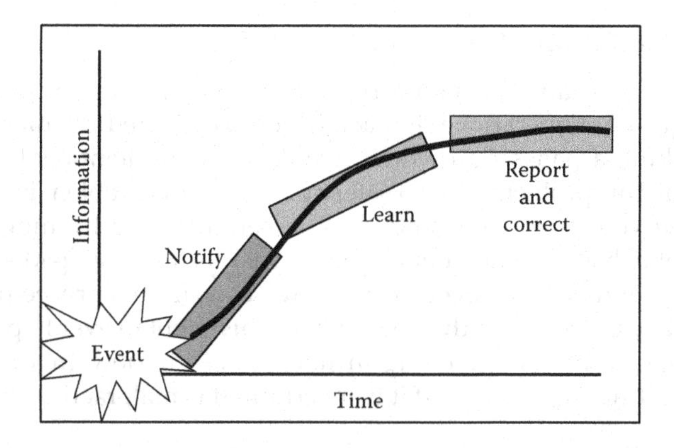


Learning is a process that grows over time. The learning process relies heavily on the people who are participating in the learning activity. It comes from the social systems – workers know how work is being done, the way the investigation has been chartered or assigned to the Learning Team (what the team thinks they are supposed to be learning), and the organizational power that is present (or assumed present) within the Learning Team (who cares about what they find out). There is a point during the learning process where enormous amounts of information are gathered. There is then a point where learning has reached a critical mass and the amount of new information slows down – the team seems to be learning the same things repeatedly. It is at this time that the learning is nearly as complete as it can be and an understanding of how the event happened emerges. The test is relatively straightforward, "can you explain how this event happened?" If you can describe the event, you have as good of a learning product as you can get. If you cannot explain the event, the learning is not complete.

That cycle seems to be present in all our operational learning activity. So many organizations stop before the critical mass of learning happens. So many organizations only stop too early in the learning process. To be brutally honest, many organizations don't do much learning at all. Those organizations just start fixing problems based upon what they think has happened.

In fact, it is not uncommon to hear managers talk about not really knowing how the event could have happened. When you hear this, you know that the pressure to find and a fix has completely consumed the time needed to explain and understand. Building an understanding of the importance of the learning in the event is paramount to our successes; in many ways, the absence of learning could be the missing component to our next level of success.

## *Fork truck versus pedestrian*

Hitting a pedestrian with a fork truck can be deadly. Even a close call can have a significant operational impact. It would be incredibly rare for a fork truck striking a pedestrian worker would ever be intentional. If it were intentional, this problem is not a safety problem; this problem is a criminal problem. If you have done good work in selection and training workers, you probably have weeded out all the murderers and sociopaths. Treating a fork truck and pedestrian event as if it were a criminal proceeding is neither effective in changing the potential for this event to ever happen again nor is it incredibly enlightening in understanding how this event happened. Treating this event as if it were criminal is malpractice and wrong.

## *Discipline and learning*

Let's start this discussion with what is sometimes confrontational and controversial. I cannot think of an industrial accident or event that was not criminal that discipline made sense to use as an organizational response. Discipline is not relevant after something bad happens. Discipline does not have the ability to change what happened, it does not alter history. Discipline has little power in preventing the next event from occurring; it does not have the power over the future. Discipline seems to serve two purposes; both have substantial, long-term value to operations,

1. It gives us somebody to blame,
2. It provides us with a quick reason why something bad happened.

Neither reason 1 nor reason 2 has any real value outside of emotional satisfaction. Neither reason makes your organization safer and is not logical of useful.

If possible, try to get the urge to discipline workers who have had bad outcomes out of your organization's operational vocabulary. I know it is hard to do. We have been trained for years to do just that … punish. I know there is pressure, both real and perceived, to make sure that we hold somebody accountable for what happened. I understand this. I get it. Stop yelling at me.

I also know that the need for discipline really screws up our ability to understand and learn. Discipline is the opposite of learning. To me, the need to learn far outweighs the need to blame and punish, but to other safety professionals and leaders that need to discipline must be strong. So many people get so angry with me when I have these discussions about discipline. I don't know when I started seeing discipline differently. I just know I came to realize that all the work we do to make people "pay" for what had happened in our organizations does not seem to make our organizations safer. And, I have a lot of quantifiable data, years of experience, and many managers who have changed, that proves that point.

I guess my response to the use of discipline would sound something like this, "We have been using the discipline model to manage safety learning and investigations for many years. Has that model made us safer or smarter about preventing the next accident?" In other words, how has it been working so far? When I go to an organization that has recently suffered some type of operational upset, that one question is often enough to start a new and different conversation.

## *Discipline is never an appropriate response to an accident*

If an accident is an accident, an unexpected combination of typical performance variables, then an accident is not intentional. If an accident is not intentional, then punishing the worker who accidently did something they did not intend to do is not an effective way to make your organization safer.

## *The test for the proper use of discipline as a safety management tool*

This is really less of a test and more of a promise to the workforce. It says before we use discipline, a valuable management tool to be sure, we will make sure that the following questions have been asked and answered. In reality, these questions guide the organizational leadership towards a more enlightened approach to worker accountability. See if you can live with these questions before you use "the stick." The questions are not difficult, just deeply inward looking for an organization to better understand how leadership is feeling about the event.

1. Was the learning event done before the discipline?
2. Would discipline ordinarily be issued (based on this behavior) even if there were not an incident?
3. Was the work activity bypassing a routine safety procedure (without an approved workaround document)?
4. Was the situation complicated enough to require adaptation?
5. Is the discipline being considered consistent with past practice? Is the previous practice now appropriate?
6. Was the action a willful, intentional violation?

If the answer to any of these questions is a "no", then the organization promises its workforce that discipline will not be used as a corrective action for this event. The transverse of this is the simple fact that if the answer to any of these questions is "yes" then discipline should be used as a corrective action for this event. Beware that this seems overly simple, but in reality is very, very effective for this organization. In many ways (and according to this company – a very impressive company with incredible safety and production numbers – this checklist has made their managers smarter and better leaders. This same set of questions also has had a dramatically positive effect on the organization's culture for safety, quality, operations, and production.

# chapter ten

# Learning teams

> Why should I clutter my mind with general information when I have people around me who can supply any knowledge I need?
>
> **Henry Ford**

As organizations begin to shift their thinking from the old view to the new view of understanding and defining safety, the journey is an interesting one. If the work is done early with leaders and decision makers, the shift in thinking seems to resonate loud and clear. The organization appears to understand that it is time to do and think differently about managing human performance. The reframing of safety as a thought experiment is sometimes painful, often controversial, and eventually truly enlightening. However, the payoff for reframing the way your organization seems to be very high.

This new way to think about how humans perform creates amazing results and engagement, the only problem is the next question: "What do we do with these ideas?" To me, that question is not that important. For many others, the question of what should I do next is vital. Managers and leaders really want a "go do" that they can immediately start working in their workplaces. The action that the managers perform is clearly not as important as the shift these managers must make in their thinking. However, the need to act is extremely strong within this population, so action will be significant.

I believe that by shifting the thinking about safety, changing the definition of what safety is, we are actually accomplishing a very aggressive intervention with the organization. Managers and leaders seem to need a bit "more meat on the bones" "then the "mental shift will create change" argument. I understand and accept that we must have some actions to follow the new ideas.

It is important to me to defend this argument a little bit more. I am talking about the discussion about change being a product of how we think differently, which in turns causes a leader to act differently. This discussion is a bit "Chicken or Egg," and can lead one into a lengthy philosophical discussion. Let us use the idea of organizational change as a way to talk about the new view shift in thinking.

Change happens one person at a time. Every conversation you have that successfully starts to build a bridge to new thinking is impactful, important and has value. Honestly, these discussions, one worker or leader at a time, are pretty much the way change happens in all groups, organizations, political parties, and eventually whole countries full of people. Change by conversation is a powerful tool. Change by conversation is how all-new philosophies emerge. As Mahatma Gandhi said, "Be the change that you wish to see in the world."

As is the conversation itself new, ideas become increasingly understandable. When you help a leader learn a new way to "frame" safety and performance leadership you have started, a new set of mental actions that influence decisions and opinions of that leader, chicken or egg? You are helping to introduce a new way to see the workplace. After all, you can never unlearn a new idea. This new value set, the new view, doesn't go away. It is more like a little flame that is constantly growing and becoming warmer. With proper care, this new flame can create great, bright light. With care and feeding, positive reinforcement, and stewardship of these new ideas – change will happen.

Managers and leaders want action. These leaders want to do something more than just think differently. Leaders want touchable differences. Just talking about change does not seem to be enough, it is enough, but there must be more that can be done. Something they can see and measure. Something that they can change and do differently – and in doing something different will lead to new and better outcomes.

We have discussed the dangers of seeing safety as an outcome. We know the safety, as an outcome attitude is a sign of a leader who clearly needs more conversation and time to think about the new view. Therefore, we teach more and try to not become defensive when we get pushback. Understanding the entire time, it is about the conversation we have with these leaders. This is not a game to be won, this new view of safety. These conversations are about teaching, not arguing. Teach every time a chance arises to talk about another way to think about safe and reliable operations. Don't react to the leaders pushback by trying to argue the right and wrong of the differing approaches instead deliberately respond in a way that helps build an intellectual bridge to this new approach to safety learning. Remember, we have to meet the leadership of our organization from where these leaders are on this new thinking, not where we wish they were. In addition, don't let it get personal. Keep the conversation away from emotion and keep it logical.

The bottom line is that you must have something tangible that you can do to buy some time and space to continue the educational discussions. That need for a tangible "go do" has given birth (and a great sense of importance) to a learning tool that does two things: is actionable and measurable and reinforces new safety thinking.

That learning tool is the Learning Team. Knowing that workers have operational information that managers and leaders, planners and engineers, don't have indicates this tool was developed to improve our opportunity to gather better, more accurate operational wisdom. This tool allows a manager to have some type of new response to an event or narrow escape. Learning Teams are different enough to appear to be something new done in response to a failure.

A word of caution seems fitting here; this is a tool to increase worker engagement and improve operational learning. It is just one tool among many tools, it is a good tool, a great tool, but it is just a tool. It can't be and shouldn't be your entire new safety program. Team-based learning became an attractive first tool to use while the thinking is new and the excitement is high. A Learning Team helps a manager just starting this journey to have an enlightened way to do something. The benefit is this tool will lead to a completely new understanding of how safety learning can and should be happening within the organization. New knowledge is a new power. You are going to nurture this process.

## What is a learning team?

Let's start this discussion with why started calling Learning Teams–"Learning Team." In my home organization, we were in the midst of a spike in our reportable numbers and everyone was getting pretty tense. Everybody was freaking out and worried that the entire organization was going down a dangerous path. The safety management team of my organization was having an ad hoc meeting at the end of a long day. All of us were standing in the hallway in front of the division office door. My boss turned to me and said something to the effect "our people are all smart and they pride themselves in their ability to solve difficult problems. I work at a facility with many smart people who pretty much solve complicated scientific and technical problems all day long for a living. This is a difficult problem. Why don't we have these same people who solve all these complicated problems help us solve this particular set of problems that are actually having the problems?" At that moment, a light bulb went off, angels sang, the air smelled like chocolate chip cookies and we had a path forward. We started to gather some small, ad hoc groups of workers that were brought together to learn what was happening. Why were we beginning to have a run of events? What didn't we know? What were we doing right? What should we stop doing? What do we need to learn about the potential for something bad to happen to this organization?

Amazingly enough, these smart people had some incredibly intelligent answers. They had never shared their knowledge with us because we had not spent much time asking for their opinions. In fact, we had never actually asked them what we needed to do to make safety better for the facility. It was as if we didn't want their point of view represented, not because we didn't like or trust our workers, but because we thought we had to provide the answers. All of a sudden, it was looking like we were leaving the most important people in the conversation out of the conversation (and we were). We thought we knew more about their work and work problems then they knew about their work and their work problems. To make matters worse, our workers thought we knew more about how to make their work safe then they knew how to make their jobs safe. In a way, our workers had given the power of safety to the safety people, when, in fact, the most powerful people for making work safer are the people who do the work.

Boy, were we all wrong.

What started out, as a desperate attempt by a senior management team to break some cycle of events, soon had become a favorite tool used in understanding, diagnosing, and correcting all sorts of problems? It was the craziest thing; we asked workers to help us learn and improve and the workers started helping us learn and improve. Who would have ever guessed improvement would be the outcome?

We had done a lot of team efforts and programs before. We had all kinds of teams that met for all sorts of reasons, but with those teams, we usually gave them a problem to solve. Now we were asking the teams to not only help solve the problems but more importantly define the problem that they need to resolve. In some way asking the teams to be a part of problem identification made the process different for the workers and the leaders. It is also worth noting that most of the safety teams we had amassed, existed at the organizational level and served as advisors to management. These new teams seemed to exist at the worker level and served as a direct connection to operational learning and improvement.

We started with one Learning Team and ended with many, many Learning Teams. Workers were being asked to identify problems and recommend solutions based entirely on how they knew the work was being completed. The tool begins to spread throughout the organizations as an immediate step to an event response, an excellent investigation tool, and a way to get better, deeper operational information and to get it fast.

Perhaps even more interesting was the way this same team-based data collection process eventually morphed into an event prevention tool. Our folks were so empowered by being asked what they thought about working reliably in our organization that they became engaged in using their perspective to help prevent other, larger and more serious events. Now we were using teams to prevent problems before the problems happened. It was hard to measure something that did not occur, but the feeling among the workers was that this work was preventive and valuable.

That little experiment born of desperation and a hallway conversation was a phenomenal success. Our management team began to understand the importance of enlarging the discussion about safety, but in a much different way. We were using workers to do problem identification and problem solution, and best yet, we were using the same workers to do both tasks.

# *chapter eleven*

# *A phased approach to learning teams*

> It is not that I'm so smart. But I stay with the questions much longer.
>
> **Albert Einstein**

Much of the rest of this book is a discussion of how Learning Teams are structured and function and how to phase them into your organization. How Learning Teams happen; What a Learning Team does. This information is based on the experiences of the organization that have tried this deeper, worker-focused problem identification and problem solution method for performance improvement.

We purposely keep the learning process simple and non-threatening so that we can learn complicated and complete information about our operations. In a world that is perfect and happy the only structure your team would need would be a very simple question, "something has happened, so what should we learn?" That is precisely the right question for your teams to tackle, it may be a bit to "free flowing" for a team of highly technical workers to grasp and use. We did not put together a procedure for a Learning Team, a qualification for Learning Teams or a reporting structure for Learning Teams. After a tremendous amount of internal struggle, the decision was made to allow the Learning Team process to be very organic and unstructured. The lack of structure is often difficult for organizations to both understand and process. But, the lack of structure also made it possible for learning to happen without a lot of expectations from leaders, supervisors, and the more detailed-driven part of any organization. In short, we allowed the learning method to develop within the context of the problem. We did not attempt to force function a problem-solving method onto the team.

That meant that we had to formulate enough structure to create comfort for a technical team to use. Something without so much structure that the question of what we should be doing merely forces the team to take the team learning activity in the direction that we have always taken before. We worked hard to make these teams seem different and more immediate than other team activities had in the past.

Over time, and after many small experiments, the method that we will introduce will begin to work well with your groups of workers. And, it will be loose enough to give ample space for learning while being structured enough to feel as if the team is following a process that has an expected outcome and an ending. This suggestion of a path forward is an excellent way to help structure a team to be successful in operational learning and understanding. The structure for a Learning Team, as introduced in this book is, however, by no means the final answer to how learning happens. In fact, this is simply the product of doing thousands of Learning Teams and figuring out what works best to keep the team moving forward and making a positive contribution to the organization. Learning Teams are fewer formal methods of collecting information from workers; any method that helps you accomplish that mission is much better than just making the information up.

I am completely convinced there is no final best way for learning to happen. I have done many, many teams and have never had two teams that are alike. In a way, it must be like doing investigations. I stopped long ago telling people how to do investigations. I have never done an investigation the same way twice in my entire career. I can't teach you how to do investigations, but I can teach you how to "think" about doing investigations.

This same idea holds true for Learning Teams. I can't teach you how to do a Learning Team. I can teach you and your team members how to think about a Learning Team. However, since, I have found that organizations just starting this journey need some sense of structure to help them succeed. Telling organizations to just think about learning is not active enough, nor substantive enough, for an organization that truly desires to improve their performance by getting smarter about their operations.

This process is not the only way, and different teams learn in various ways. There is no perfect method for learning. You are constantly re-creating learning environments in order to get better, newer, and more efficient understanding of your operations. What is really happening is that your organization's operators and managers are gradually getting wiser and wiser about how work is actually completed. In gaining operational wisdom, the outcome is better – more informed – operations.

You will notice that the word "phase" is used to describe each part of the learning structure. I found this terrific description from Indiana University School of Engineering a while ago on how to illustrate the importance of using the word phase. According to the Indiana University description, a Phase denotes the particular point in the cycle of a waveform, measured as an angle in degrees. It is normally not an audible characteristic of a single wave (but can be when we are using very low-frequency waves as controls in synthesis). The phase is an imperative factor in the interaction of one wave with another. These learning phases are not steps to be followed by robotic compliance; this is not a recipe or instructions for making teams learn better. These learning phases are interactions – learning waveforms – of many perspectives of how work happens or happened. The idea a Learning Team would occur in a linear, step-like fashion is incredibly restrictive and structured with an idea to use in the process of creating a Learning Team. Using a linear process to understand, guide, or even describe a non-linear event is a quick move right back to the old-school way of doing operational learning. We want the world to be linear, so it is understood we would want learning to be linear as well.

We need to build a method for learning that understands and speaks to the space between the worker doing work and the work that is done, the place where the worker meets the work. This worker/workspace defies using linear explanation because it is not a direct activity. Many conditions exist at once in the process of doing a task. For the most part, the process of doing work is actually a complex operation. The only way to talk about the complex relationships that exists between the worker and work process is to use a process that allows the discussions about these complexities in operations. Having conversations about learning enables us to make the work process less obscure and mysterious and more transparent and understandable.

Learning happens in phases. The traditional and historical use of the word "phase" is work that is carried out in gradual stages. Think of these phases of learning like a light that gets slowly brighter, as this light becomes brighter and brighter, more and more of our organization's operations become clearer and clearer. It is very similar to the way the managers of a movie theater control the lighting of the movie auditorium. Slowly and slowly the lights dim before the film begins – creating the mysterious dark room for the movie. When the movie is over, the theater manager gradually brings the house lights up revealing over a period of time the dingy and popcorn filled room in the movie theater.

Gradual stages of knowledge are precisely how we want a Learning Team to understand and explain an event. Slowly and surely, we want to turn the lights up in our operations. The brighter the lighting becomes, the more the organization will know about the workplace and the way the work takes place. Create a team that gradually becomes wiser and you will have created a team that changes the performance in your organization. Remember that it is the process. It is more about the journey than the destination.

This gradual wisdom seems to happen in seven phases. These phases are additive to one another. No one step is sufficient to give your organization a complete understanding of how operations occur. That said these phases might not all be necessary for learning to happen. This list of learning phases is more of the guide based on past successful experience, not a strict list of activities to complete. The team will find the right balance for using these phases. Better yet, I predict you will get excellent at navigating these phases of learning with your teams. So good, in fact, that you will eventually find yourself less and less a part of doing these teams and more and more the beneficiary of the information that the team learns.

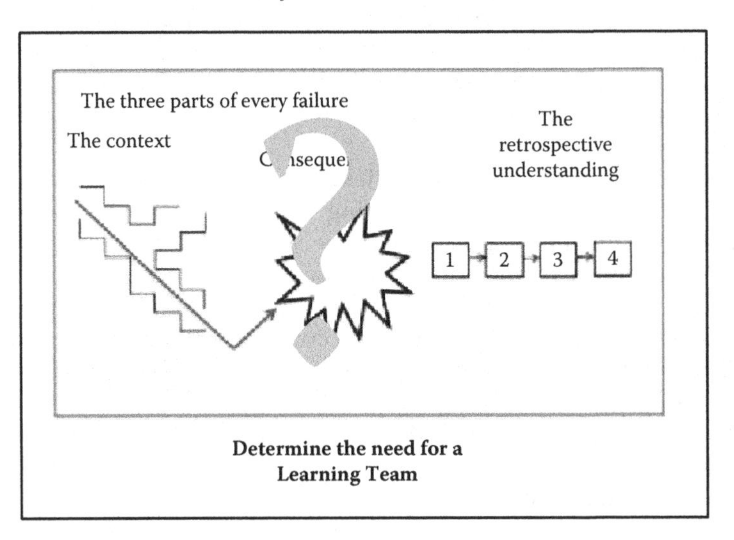


**Determine the need for a Learning Team**

## The Learning Team Phases

**Phase 1 – Determine need for Learning Team**
Phase 2 – 1st Session – Learning Mode only
Phase 3 – Provide "Soak Time"
Phase 4 – 2nd Session – Start in Learning Mode
Phase 5 – Define current defenses/build new ones
Phase 6 – Tracking actions & criteria for closure
Phase 7 – Communicate to other applicable areas

# *chapter twelve*

# *When to learn?*

## *Phase one: Determine the need for a learning team*

> Tell me and I forget, teach me and I may remember, involve me and I learn.
>
> **Benjamin Franklin**

### *Determine when to learn*

When is an event an event worth investigating? The short is every event is information rich, filled with operational intelligence, as is the pre-event prevention stories. The problem is you can neither afford nor have the time to learn from every single event that happens in your organization. It is incredibly difficult to predict which events are going to be the catalyst for the best organizational change. Therefore, you need some practice.

That said, some of the events that you will learn from are quite apparent after some type of event has happened. When do you need to know more? The quick answer is when don't you need to know more; in your world you can't know too much, about how your operations happen. After all, you don't know what you don't know. Wisdom about your people and processes is helpful and valuable. Operational wisdom is the difference between fixing the right things the first time or fixing the wrong things aggressively and often.

Yet, it would be impossible to learn everything and not everything needs to be learned. You have limited time, resources, and energy and you must use your time smartly and efficiently. Let's face it, not everything that happens or could possibly happen is operationally attractive. You can't stop production and solely spend your time in learning mode.

How do you know what is important? Sadly, the answer is you probably don't know what is important and what may be less significant in your operations. Every event is information rich – your job is to find out when it is time to learn more about the event. The test best used to see if you know enough is quite straightforward: Any time you can't explain how an event happened (or could have happened), you don't know enough about the particular event. If you can't say why a smart worker did something unpredictable, unexpected, or operationally strange, you don't know what happened and you have not learned enough.

Using a small Learning Team to do the first pass on event learning is an incredibly efficient way to determine if there is anything there. A small team brought together quickly after something is discovered is operationally smart and administratively agile. A small Learning Team will help your organization determine what is going on operationally.

Most importantly is the power that asking workers to help identify the problem create ineffective engagement of workers. By asking workers to help identify the problem, your organization is engaging your workforce in the process of using their knowledge for the benefit of the organization. This form of engagement is imperative for the creation of problem ownership. This power is even more amplified when you also include your workers in problem solution – but that comes a bit later ... for now let's stay in problem discovery.

The process of putting together a team to discover what is happening in your workplace is enormously empowering to the members of the team. In many ways, that sense of empowerment makes all the difference. Not only does the team learn about events and pre-events, but they also become empowered to take this knowledge even farther in to your organization's operations. That ownership is impressively powerful and useful.

## *Questions to start your thinking about the learning process*

Here are some questions that we use to build a case for learning:

- Can you describe and explain how a typically smart worker found himself or herself in the situation where the outcome that happened (or could have happened), occurred? If not, learning is needed and warranted.
- The mishap (or near mishap) happened. What benefit might we gain from understanding the context of this work situation before the mishap appeared?
- Do you see a series of weak signal events that could indicate a drift from your regular operations? Does your system seem unstable?
- Is the current situation or event surprising to you, your workers, and your organization?
- What is our best procedure? What is our worst procedure?
- Where will the next accident happen and how?

- Does what happened make sense to you, given your knowledge and experience with your organization?
- Are you lying awake at night? Do you really know how work is being done?
- Did you successfully deliver some outcome that stretched your organization's capacity? How did you do this job so well?
- What do I need to know about our operations?

This list is only partial; it would be very difficult to ever represent all the potential questions you could ask about your operations. But the general theme of all these questions is this: Do we need to know more before we act?

The problem is that these issues require courage. Asking questions that almost certainly are going to require more of your time, resources, and attention is hard to do, not easy. The bravest leaders in the world are "less than comfortable" with hearing that there are complexities affecting both production and behavior existing in their organizations.

Remember, the first, best hope you have for reducing the frequency and severity of events that happen in your organization is wisdom about your organization. Brave as you have to be, the payoff of knowing more clearly outweighs not knowing. You simply have to know. Knowing early gives you much more response time and space. It's a bit overwhelming at times, but knowledge is vital.

Phase one decision-making should not be difficult; all you are doing is temporarily sending highly trained and deeply experienced system experts into the field to be your eyes and ears. These workers, the masters-of-how-work-is-done, will tell you very quickly what it is you need to know. When in doubt, send in a quick team.

## *Don't start with problems that are too big or complex for the group*

A big problem is never just one big problem. Significant problems are always a collection of smaller problems that appear directly intertwined. Operationally you don't ever have big failures; your operation probably has had or will have a series of operational variability that collectively appear on your radar screen as one big problem. You don't have big problems – you have many smaller problems that give the appearance of a more significant problem. The power of operational learning is in letting the Learning Team set the boundaries for how much of the problem they feel they have both the knowledge of and the potential to … analyze, learn and change. Solving many small problems leads you to the solution for your larger issues. Thinking small is powerful in the collective nature of team learning.

Allow the team to start the problem identification process with a blank sheet of paper, and then trust your workers will find improvement topics that will be the beginning of much larger improvements. Sometimes that means, or seems like you are giving up a bit of control. It is a leap of faith, but with proper guidance and honest intentions, the group will soon understand their job is to create good in the world and your organization.

Teams will understand that small problems are more workable without speaking directly about making the problem smaller. Allow the team to understand the greater impact of the problem and then decide what part of the problem they should tackle first. Learning is about having many small wins that will equal larger wisdom, I promise.

Have your teams out-brief you about their progress if you feel you need more control over this process. During your briefing, I would caution you that your job is entirely to nod, agree, encourage and ask them to dig even deeper. Don't correct, add, limit, or delete – at least not during the selection and problem identification phase – let the process work and see if you are surprised by what you learn from your workers.

Ensure learning success by helping your workers identify and solve small problems. Know that these first steps are how your journey to improvement will create operational knowledge. Fix the small problems and the large problems will take care of themselves. Need proof? Check out the "broken window pane" theory and apply that idea to your operations. This theory, quickly summarized, states that if you take care of the little problems the big problems will never happen. If you own a building you should replace the broken windows as soon as possible after the windows have been vandalized. This constant, immediate replacement of the broken window panes communicates to the vandals that you are constantly watching and monitoring the building which will decrease the number of time you have to replace vandalized windows. Early detection and action of small-signal events helps to decrease the number of large events.

## *Safety only?*

One of the surprising outcomes of using workers to identify problems and solutions in work as it actually happens is that the team problem-solving techniques will be used for other problems within your organization. By empowering and engaging workers in this way, you build skills in improving your systems and processes.

Those skills don't lie dormant once they have been discovered. Your workers will use this process because they believe you support and approve of this type of problem identification, and more importantly because the workers know that this problem identification and solution process works for them. You are giving a voice to the workers using the language of process improvement.

The metric I am always excited to identify when I go into an organization is the presence of flip charts in the production area. It is even more exciting when you realize that you are looking at notes from a production problem, a quality problem, and an ergonomic issue. This is only made better when the workers show you the solutions. You will never hear a smarter worker than a worker who is telling you about a problem that they helped correct.

Any operational problem (or near problem) that you need to understand better is an excellent candidate for learning. Safety issues are simply the area in which I work. I also will submit that the payoff for fixing a safety issue is a life saved or an injury prevented. Operational learning also underrepresents safety. The quality folks have been using learning tools for years. The lean people never do anything without learning first. Safety has not traditionally been given the benefit of using analysis to improve. Safety is always under pressure to fix immediately, learn later.

## *The team learns about the event, together*

Make a space for learning. I know this phase of learning, the part where you create a learning space, sounds soft and wooly. I have struggled with a better way to communicate this idea. I know it can sometimes be hard for a technical person to be comfortable with the most personal part of the work. I can't find words that are better than the concept of "making a space" in your organization. I can promise you that a critical success factor in creating change in your safety program is to create a space, both actually and psychologically, where learning can happen. When you give your workers space to learn, coupled with an expectation of some type of learning outcome, learning will transpire.

## *Setting the stage for learning*

Creating the environment where people in your organization can learn is important, critical. Workers need a place where they can take notes, hear each other, sit and rest their weary bones, and psychologically break away from the performance of work.

Workers cannot learn and perform at the same time. No one can learn and perform at the same time. Learning and performing are opposite activities. Most high school and college sports team take video of the games that these teams play in competition. These games are not recorded solely for a historical record. These games are recorded so that the team can collectively watch and review the game, dissecting the game into the smallest particles to learn and improve the sport.

The players don't really need to watch the game. They played the game, and they know exactly how the game will end. There will be no surprise endings or plot twists for this game video. What these players and coaches are doing is changing the perspective of the participants. When you are playing the game you are not analyzing the game; you can't analyze the game. When you are playing a sport in competition, you are not learning the game.

The same applies to the coaches. The coaches are actively acting and responding to the game that is happening all around them. The coaches are not actually learning about the game that is being played. These people don't have the time or the excess attention that is needed to learn. The coach is coaching.

That is exactly true for your organization. Your workers don't really have the time or the excess attention to actively analyze the work as completed. Your workers are acting and responding to the work that they discover before them. Your workers can't really learn about the work context while performing the work. This difference, the difference between learning and performing, is not controversial or even a new idea; it is the way human beings act out their lives.

Performance and learning contradict each other and that is paramount for you to know. I often talk about the creation of learning space as a secret weapon for success. The reason I talk this way is that the attention conflict between learning and performing is so true. Simply by creating psychological and physical space and time for learning to happen changes the way you receive operational feedback.

You don't need to create an actual place for learning to occur per se, but a real place for learning is a terrific idea. I work with a gigantic warehouse operation, a large warehouse that created an area in the middle of the warehouse area with good lighting, freshly painted floors, and whiteboards surrounding this space. This learning center was established to provide a space for learning to happen.

This space was not the break room; this warehouse has a nice break room that is also comfortable, relaxed and functional. This new work space was filled with flip charts, markers, tape, tables and chairs, and people understanding and identifying problems and then recommending ways to make the work better and more efficient.

The workers saw the space as an oddity at first, but before long the workers were using the space for operational improvement. Interestingly enough, the warehouse manager also eventually placed his desk adjacent to this learning area to increase his accessibility and attention to the improvement work transpiring. This warehouse changed, dramatically changed, when the workers were used in both problem identification and problem solutions. What changed was that the workers were given a voice in operations – a voice that was so compelling the boss moved his office so he could hear the improvement news louder and quicker.

## *Who should learn for your organization?*

The team should be made up of people who do the work. Worker involvement is fundamental, vital to your success. However, the most challenging individuals to pull from the production floor, out of your operations, are the workers doing the work. Nobody wants to take people out of production work to talk about a problem that has not even happened yet. Without proper preparation and management commitment, pulling a worker from a functioning production line takes guts and confidence.

The more seductive option would be to pull managers, office people, engineers, and safety people into the team because they are so much easier to access from within your organization. The management level has time to think and analyze because they are not a part of the production pressures that exist in the organization. The management layer is easy to access. Workers are hard to access. Don't do it. Don't be seduced by what is easy and non-impactful to operations.

Your teams should have somewhere between four and six members. More than six is hard to manage, big groups are hard to brainstorm with and sometimes scheduling can get confusing and complicated. Fewer than four limits the pool of information from which your team will access and discuss. There are certainly many exceptions to this rule and your teams will shrink and grow as needed. You will clearly identify people who should be on the team who are not represented. As you learn more about the work transpiring, new discoveries will lead to deeper and different experiences being represented. That's normal and is a good sign that learning is happening.

Mix personalities and expertise. If you have some quiet people on the team, be sure to balance the team with some more outgoing workers as well. Choose workers who will exceed in this activity, people who like to do this type of group activity. Set the team up to be successful and the team will succeed.

## *Does the team need a leader and/or coach?*

The quick answer is yes and no. I realize that is about the cheesiest way to respond to a question possible. Worse yet, I asked the question. Teams without leadership will quickly develop leadership. Leadership will emerge as a part of the Learning Team, but does the team need some guidance as the leadership emergence process happens?

Here's what we have found ... you will have a leader appear and help guide the team. However, a coach placed to start the process dramatically improves the team's success. The coach is not the leader, but more of a timekeeper, scribe, room arranger, and initial guide in the process.

You have seen these words before in this book; the goal is to keep the learning process as simple and uncomplicated as possible so that it is easy to learn complicated information. Having a coach (or coaching) to help the team get started removes some of the uncertainty out of the process and speeds up the process of the team forming on its own.

The coach role should not be complicated, difficult, or elite. It is simply a person who can say, "go." They then make sure the knowledge that is being discussed is being captured in some way so that it can be referred to and used to analyze the work as it is being done.

### *Does phase one matter?*

Yes, determining the need for learning matters. Do you need to learn? Probably yes. Do you have experts in your organization? Yes, very smart ones. Can they teach you something you did not know about your organization? Absolutely. Is your ability to understand more about how work happens in your organization critical to your success and the success of your organization? That is why we are here. That new knowledge is the missing part from our current methods. Knowing more makes us better.

However, it only works if you decide you have a reason to learn. If I may be so bold, you have many reasons to learn. Your perception of work is never as accurate as the understanding that lives within the people who do the job.

My advice is trying a Learning Team. The most powerful tool you have to facilitate change and improve performance is the small experiment. You always have the ability to test your learning process by actually doing a test of your learning process. The next time you have an operational curiosity, something happens that you cannot explain or understand, bring together some workers and ask them what you should know.

### *Summary of phase one*

Learning Teams can be used to understand any operational function. Learning Teams are often ad hoc and appear as the need immediately arises. Using a Learning Team as an immediate action after some type of operational surprise both increases understanding and creates knowledge and stability. Teams exist to learn about how work is done. Any problem that is not clearly understood benefits from more clarification – a Learning Team.

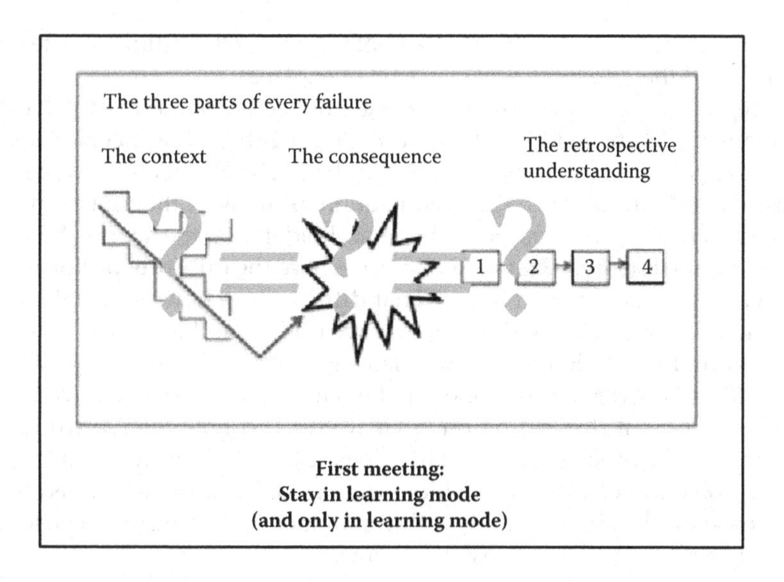


**First meeting:**
**Stay in learning mode**
**(and only in learning mode)**

**The Learning Team Phases**

Phase 1 – Determine need for Learning Team
**Phase 2 – 1st Session – Learning Mode only**
Phase 3 – Provide "Soak Time"
Phase 4 – 2nd Session – Start in Learning Mode
Phase 5 – Define current defenses/build new ones
Phase 6 – Tracking actions & criteria for closure
Phase 7 – Communicate to other applicable areas

## *Phase two: The first meeting – Discovery*

You have determined you have a need to learn about some event or a near event. You have an idea of whom you want to be represented on the team. You have created a space for learning to happen and performing to stop. Now, it is time to have the first session of a Learning Team meeting. Pretty exciting stuff, isn't it? Hold on before you get too excited. This first session is important, but it will not be giving you any solutions. Don't expect the first session to give you answers to your operational questions, not just yet. You are just starting the discovery phase, and it is important to re-emphasise that the need for a solution will only get in the way of the learning. Make sure that the expectation of the first meeting is that it is process and that all answers won't be identified or resolved after the first meeting.

Goal setting and expectations must be established early in this session. Teams normally get together to either solve a problem or create some type of value. A Learning Team does both, quite well. Getting your team to understand their purpose will take a bit of explanation.

Begin with a discussion of how the knowledge about the work exists in many different places within your organization. Engineers see your organization as a system of drawings and equations that create some type of product. Managers see the organization as an asset that makes money or adds value to the company, its shareholders, or the world. Workers, however, see the organization as a place where the dutifully perform their individual task to make a living. That difference in how we see the organization changes what we look at, how we understand the work, and what we eventually will do to make problems go away.

Although everybody works at the same place, the place where we work is different depending on what we do and how we see work. The position within the facility actually does influence how we identify and solve problems because our positions define our point of view. This changing sense of identity is not a bad thing; in fact, for learning purposes the identity differences are probably beneficial.

This discussion both engages the worker and establishes expertise, broad knowledge, and credibility. The worker is the expert. The worker sees the organization in a different way and, therefore, the worker will help the organization understand the work differently. The worker is the only person in the organization that has this particular point of view, and the worker by definition is the only person in the organization that can have this point of view. This discussion is valuable.

What is happening is that you are building both confidence and capacity for the Learning Team. That sweet combination, confidence, and capacity, is just the fuel you will need to discover how work is done or how a mishap could happen. The workers have a unique point of view. The workers are the only group that has that point of view. The workers are being asked to talk about the work from that point of view.

Charge the team to help the organization get wiser. Ask the team to tell you what you should learn. Absolutely ask the team to help you solve this problem. Admit to the group that you don't know; not knowing is actually a powerful position for you (and the team) to take in this discussion. Not knowing means that you have the potential to find out more. Knowing more is powerful. Engage the team in helping define the problem and design a solution for that problem.

Once the team is identified and the learning need has been established, the work of discovery begins. Discovery is a bit tricky. It is hard to stay focused on problem identification. You are not good at it. I am not good at it. We find the analysis phase of learning inactive and stagnate. In fact, learning is not nearly as exciting as doing. The reason for that lack of excitement is the lack of action. Let's get one thing straight: Learning is a corrective measure. Learning has value. Learning makes us smarter and better informed – wiser.

Learning is vital and primary to fixing. Learning is doing something. Learning is a form of action. Learning saves lives and money. Repeat that mantra over and over and over again. This is important and I will next tell you why. Learning is action. Learning is action. Learning IS action.

## *Stay in problem-solving mode*

The biggest problem you will have is keeping the group of motivated and informed workers in the learning mode. Good workers fix problems. Good workers have solutions for problems. Good workers don't bring you problems – they bring you solutions. All that action by good workers is a problem, a significant problem. Quite often, organizations tell workers to "just fix the problem" and not bother anyone with it. The group will want to fix something. Getting and keeping the group in discovery mode is a challenge you will have to solve before you can go much further on this journey.

The need to stop learning and start fixing is very strong. So strong, in fact, that this need to solve the problem will overtake your analysis of the problem. Be acutely aware that the team will quickly drift towards fixing the problems. You must help the team get beyond the need to adjust and remain in learning mode. This seems simple (and quite honestly over-discussed in this section) but staying in learning is not simple; it is quite difficult and takes attention and discipline.

Learning Teams are actually taking tacit knowledge and transforming it into explicit knowledge. We are trying, as simply, and as uncomplicated as we can, to understand and explain how work happens from the workers point of view as entirely as we possibly can, in as much detail as practical. Better still, we are going to try to accomplish broad discovery in only a couple of hours – three at the most. This all sounds pretty complicated, but actually, this is quite simple and elegant.

Once you establish a learning environment, the team will stay in discovery for as long as the team feels the need to remain in this phase. You will collect a lot of information, many work conditions that are present in the work. That is OK; in fact, this is precisely what the team must do. Don't spend time editing or arguing, just discover and record.

Every Learning Team should probably meet twice at a minimum. There is a reason for this somewhat artificial division and introduction of a second meeting. The reason is: You need the second meeting so you can keep the first session in the learning mode. The addition of the second session gives you some structural discipline. During this first session, you can actually tell the team that we will only be learning about the problem; we will not be solving this problem during this first meeting. Your people will want to address and solve the problem during that first meeting. Are we not taught that action is always better than inaction? In the case of problem identification, the answer is no. You must have a period of inaction – loosely translated. When we jump to solving the problems – we stop the learning process. Managers get "fix-y" and that is not always a good thing. It is natural and normal, but not good.

The enemy of the question is the answer.

When you think you know the problem, you stop the process of discovery. In fact, when you move to the solution phase you move out of the discovery phase, and the minute you move towards solving the problem you are no longer trying to fully understand and explain the problem. It is the end of the learning process. In many ways this is a rather extreme version of confirmation bias. As soon as we see what we think we will see, we stop looking for anything else.

Setting our need to immediately fix the problem aside for a moment, let's discuss another human bias that impedes our ability to understand and explain how mishaps or near mishaps, happen. This problem is a massive shift in thinking, but hold on before you get too concerned. There is an alternative to our traditional methods and tools.

## *Start a loose representation of the event or of the work in question*

Timelines are bad. Timelines are the way every investigation that has ever been done throughout the history of mankind (Okay, I made that last part up). Timelines make sense if you think of the way human beings think about events – in retrospect. We give events order and importance after the event has taken place. The problem is that timelines bias our thinking in two important ways: Timelines assume order and timelines assume cause. Much more discussion of this interesting notion is below. This bias creates a set of circumstances in retrospect/discovery that just did not exist while the work (or event) was happening. Saying timelines are wrong is pretty serious, because we are criticizing the most fundamental of all event-learning tools that we have used. Most organizations don't even consider a timeline a dangerous thing. In fact, most organizations see a timeline as factual and objective. Timelines are not accurate (they are created in retrospect) and they are very biased towards linear failure modeling. If given the chance I would never-ever use a classic timeline in any investigation. Timelines are really that bad.

The two imperfections of using a timeline to represent an event:

1. Timelines assume everything happens in order. This happened first. That happened next. The third thing happened after the second and on and on. Timelines are a neat and clean way to give the event order. Timelines allow us to make sense of a particular time-period where something happened. The problem is the event did not

have that same precise and linear order when the event occurred. If you look at an event, look deeply, you will find that many of the events that seem orderly in retrospect all happened at the very same time. We give events order. We make up timelines to discuss and capture super complex events in a simple and clear fashion. That clinical order we place on an event in retrospect is both unfair to the workers who were involved in the event and it dramatically simplifies and clarifies the way we have imagined the problem happening. The timeline for the event in question did not exist before the event transpired. In fact, the timeline for an event is constructed from our impression of what happened.

2. We usually start at the end of the mishap, the consequence or adverse outcome, and construct the timeline backward towards the beginning of what we think of as the event. It is an excellent way to build a timeline, but it is a horrible way to describe and explain how the event happened. What we are doing when we start at the bad outcome and build a timeline backward is identifying the bad thing and then we carefully and thoughtfully look backward for what we think could have caused the bad thing. Our intention is pure, but our logic is flawed. We are primarily making up what we believe could have created the bad thing. Ultimately, this thinking will drive any investigation or Learning Team to the worker who was closest to the failure. The most recognized example of this is a problem in almost every organization is best summarized as, "the last person to touch it screwed it up." That logic, often called the last person logic, is incomplete, oversimplified and usually wrong. Sorry, but it really is wrong. It's also mean and unfair; in many ways, you are making the person who already feels terrible about the bad outcome seem even worse.
3. Timelines imply that the workers could have somehow predicted the future. When we put an event on a time line, the event looks very ordered and simple to understand. In a great way, this order makes it seem easy to predict what would happen next. In reality, we have no idea how difficult the outcome would have been at the time for the workers involved to predict. Timelines tend to imply the workers should have easily known the outcome. If easy outcomes were really easy, the unwanted outcome would have never happened.

## *So, what do we do if we can't use a timeline?*

We do have to symbolically represent what happened so we can explain and learn how the event happened and timelines are a standard and acceptable way to do just that, symbolically represent a mishap. Even worse still, the organization or the regulator practically demands a timeline, so much so that if you don't make one it is predictable that someone else will. In essence, we almost have to make some type of timeline for our mental process and the organization's ability to "code" the story.

The answer to this problem is quite straightforward. When you make you timeline representation, ask the workers to start at the beginning of the work activity or the workday and tell the story they have to tell moving towards the failure. The fundamental question I ask a Learning Team is this one: "Tell me how you guys start your shift?" However you engage the team, keep the motion moving forward, as opposed to the reconstruction of the story backward.

Its amazing what happens; the workers begin telling each other the detailed information about when they come in, what they do to get ready to do work, how work comes to them, planning, resource availability, and almost anything else they do from the beginning to the point that is, by now, operationally attractive. Before long you had captured a precise and complete understanding of what was happening before the event transpired. Often times, by the time you get deep in the understanding of how work happens you have an extremely deep understanding of the conditions that existed in the workspace before events transpired.

## *Identify conditions – Not choices*

> Don't look at what choices workers need to make differently, instead look where conditions create operational conflicts.
>
> **David Woods**

One way to begin the discussion of what happened (or what could have happened if you are in preventative mode), is to only ask the people involved what conditions were present in order for this failure to occur? Very quickly, this question will begin to generate a tremendous amount of information. Amazingly, the information gathered as conditions are more complete and are often much more attractive than a linear and systematic line of questioning.

When you ask what conditions must be present for this failure to happen, you are asking workers to discuss all the things that came together for this failure to happen. This question feels safe, is not incredibly personally or professionally threatening, and is incredibly informative. Condition identification becomes something akin to a brainstorming session about the parts of an event, not the cause of the event. Even though, you will certainly begin to have an understanding of how the event happened (the traditional cause) as you start to amass your list of event conditions.

An excellent work product for a Learning Team is a list of conditions present in the failure identified.

It is easy to talk about what was in the workspace when the failure happened, a low-risk conversation. It is not normally easy to speak of the choices and behaviors, opinions and options that were present when the failure occurred, a high-risk conversation. When your questions turn to worker motivation and decision-making, you can almost guarantee that workers will begin to feel as if they were the reason the failure happened. As soon as the worker starts to believe that they did something wrong, almost all your learning will then be focused on reinforcement of that idea, often in spite of your best intentions. There is a real difference in communication when the discussion is focused on the conditions and not on the people.

## *List the conditions present in the event context*

In a way, the first Learning Team meeting acts and feels more like a brainstorming session then an investigation or event critique session. There are no witness statements, no individual interviews, or quasi-law enforcement type fact-finding activities. If the goal of a Learning Team is to understand and explain how an event happened (or could have happened), event information gathering should be natural, detailed, and as complete as possible.

In fact, organizations that are wonderful at learning have found almost the exact opposite environment creates the best information and the most success. Instead of looking for direct and detailed causes for the event, teams spend time discussing what existed in the workplace environment, the context that was necessary for the bad outcome to become a bad outcome. There are plenty ways to discover the context and it is clear that different events will demand different approaches.

However, one simple way to deeply understand the context of the work environment is to just begin asking the workers what conditions have had to exist for the event to happen? That question is very open-ended and almost limitless; this question is an excellent way to collect profound and rich information.

Conditions are the things that exist in the context of doing work. Conditions may include but are not usually behaviors. Conditions are not choices either. Conditions are an extremely loose word used purposely to collect as much context information as possible about the work environment.

Conditions can be actions, they can be mindsets, and they will include both the regular and the unusual components of how work happens. Most importantly, conditions are all the items that the Learning Team members recognize as significant and worthy of comment. Don't look for a right answer or a wrong answer; there won't be right or wrong conditions. There are simply "conditions" neither right nor wrong and they all *may* matter. That is the tricky part, you don't know which conditions matter or don't matter. So, collect them all. You will get to process these data points latter as the Learning Team continues to analyze and understand.

As the workers identify conditions, record these conditions so you can use this data in the discovery process. Anything that the Learning Team feels is a condition is a condition. Clarify if you need to and ask for more information if you don't understand what the team is saying. Don't judge or qualify the conditions – capture them. Some conditions may not seem to make sense in the discovery phase, often are paramount in the solution and understanding phase of learning.

Be aware that you will have hundreds and hundreds of conditions identified during this first session. Have plenty of paper and suitable markers – you will need them. Not all conditions matter, but let's say half are important – no let's say only a quarter of them are essential to understanding the event context. If you get 100 conditions (and that is actually a pretty typical and predictable number of conditions to get in an hour) and only 10% have learning value, my math tells me that you have identified ten things that you can almost immediately work with and focus upon resolving. "Not a bad hour's work," says the man who has spent weeks of his life locked away in conference rooms being used as investigation headquarters, trying hard to come up with some corrective action recommendations that make sense.

## *Everyone has a perspective*

One last note on the first session of a Learning Team; actively solicit involvement as the meeting begins to gel. Ensure that all team members are included and asked to participate in the learning mode. If a person does not want to provide information, that in and of itself could be informational to the team.

Don't force members to talk, but remember these people are on this team because they do this work. Allow silence and space to appear and create an environment that welcomes information about the workers perception of how work happens. One worker's perception of what is happening may not, nor should not, be the same as the other worker's perception of work. That is the very essence of team-based learning and that essenential is necessary.

It is also quite possible that the right workers will not be in the room with this Learning Team. You will identify if other representatives should become members of this team by the discussion of the learning and conditions. If it is important to include shipping, include shipping. Does the warehouse need to be represented? How about maintenance? Do they have a perspective on how the work is actually done that is not represented? Next meeting, maintenance will be included in our discussion. Don't hesitate to form and reform as a Learning Team as information and perspective is needed. That particular point of view both from and of that knowledge is precisely why you have created a team.

Allow the workers to find their voices. The Learning Team should build confidence and capacity in your workforce that workers are the best choice to both identify and fix problems that happen in your operations. This is the confidence building part of the process. You might find it strange, but many times, I find some emotional when they finally get the opportunity to find their voice during this process. Asking workers to give their perspective is very empowering and engaging.

This first session, the discovery session is both fun and significant. You are helping to build a team of process improvement experts. Knowing that you are engaging workers in learning is unique. You have a moral and ethical duty to create the best discovery environment you can create. The benefit from all of this coaching and heavy-team-building-lifting is that your organization becomes more aware of how events happen. The advantage for you is that you know more and better operational information about your organization's context. Discovery has a very high payoff.

Best of all this first learning session is engaging, exciting, and fun. I don't often get to use the words interesting and fun when talking about an investigation process, but you will soon see. The amount of pride workers take in telling you their story is imperative to the long-term success of your operations. Improvement begins when workers have ownership over the way they get to talk about doing work. Change begins the moment the workers perceive a change; your discovery activities loudly communicate a change in the way you are managing safety and reliability.

I am always amazed while participating in Learning Teams, how different workers act and talk when you walk with them to their actual workspace and ask them to tell you how they do work. Their whole attitude changes. You will witness expertise (impressive), energy (exciting) and pride (deep understanding of what is being done) on the faces and in the voices of the workers telling you about work. It really is an incredibly special, crucial moment in time. Watch for it. Better yet, create this opportunity for both you and your workers.

## *Phase two summary*

The first meeting of a Learning Team is paramount. The first meeting sets the tone for discovery, gives voice to the workers who perform the work. It creates an environment that is open and inclusive, and identifies the conditions that are present while doing the work. Phase Two cautions teams to not attribute meaning and order to the work problem while trying to understand the problem, the Classic timeline problem. Allow the story of the conditions and context to emerge from the beginning of the work for the event in question. The challenge during phase two is to keep the team in learning and discovery mode and away from the immediate and urgent need to fix the problem or problems. There will be time to fix, but that time must follow a time for discovery.

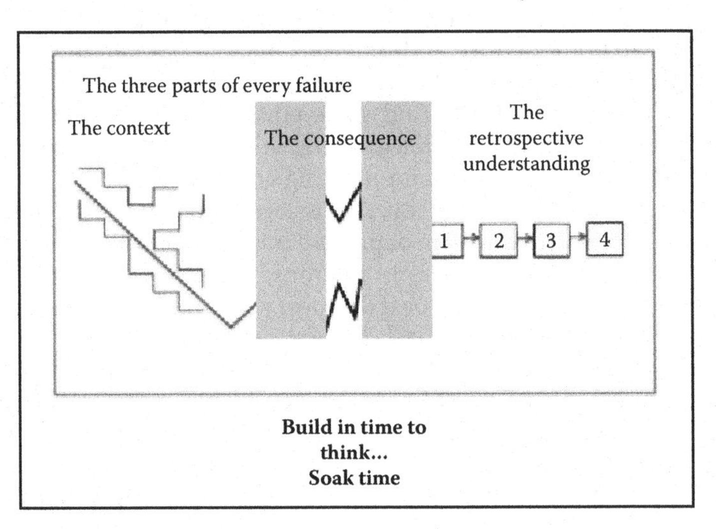


**The Learning Team Phases**

Phase 1 – Determine need for Learning Team
Phase 2 – 1st Session – Learning Mode only
**Phase 3 – Provide "Soak Time"**
Phase 4 – 2nd Session – Start in Learning Mode
Phase 5 – Define current defenses / build new ones
Phase 6 – Tracking actions & criteria for closure
Phase 7 – Communicate to other applicable areas

# chapter thirteen

# Let it marinate – Build in time to think

> It's not what you look at that matters, it's what you see.
>
> **Henry David Thoreau**

## Phase three: Soak time!

Learning doesn't happen all at once. Learning takes time and effort. So, make that part of your plan. Learning is the process of gathering and analyzing, re-gathering and re-analyzing, processing, and getting better. Learning is a feedback loop that you are naturally equipped to use – and you use it all the time. You have gathered and discovered information throughout your entire life. Every day you are a little wiser, a little smarter. Learning simply isn't fast.

One of the most interesting parts of learning is allowing the team some time to think, process and review what the team discovered. A bit of space in the development allows some room for thinking, and that additional thinking room often generates better understanding, better problem identification, and overall better outcomes. This time in the process cost nothing and adds value to this process.

Have you ever been in a meeting and been actively engaged in a discussion? Then after the meeting was over, and you were back at your desk, you thought of some crucial information that you were not able to conjure up while you were in the meeting? Of course you have, because thinking of just the right thing to say after the opportunity to say it relies on one important factor … time. A quick remark or the best solutions often appear after you have had some time to think about a comment made in a meeting, a problem, or an idea.

Thinking of the right thing to say is also a function of having some time to work through the information. It is taking the time to process the data, analysis of the overall purpose, and making some determinations of what should be added to the conversation. Just as we keep the first session in learning mode, we offer a break for people to spend a bit of time thinking about what was said in understanding how the event happened or could have happened.

This gap in the process is purposeful and effective. The gap, the soak time, creates space for the individual members of the team to think, quietly and privately, about what they know about the work. The soak time also predictably allows at least some of the Learning Team members to think of and offer additional information.

We work in a workplace where there is not much time built in to our days to ponder and think; production pressures won't allow much time for reflection. When we are in learning and discovery mode to identify and solve a problem, we do have the ability to build in processing time. There are many ways to allow your workers to think about the problems they are attempting to identify, but none seems effective as simply scheduling a second part of the meeting at a later date. It is easy and remarkably effective. The value that the team receives is the benefit of time for thinking about the problem, and usually that thinking is happening where the work takes place.

Creating space for time is actually creating space for thinking. Most workers are good thinkers. The problems usually arise because the worker stops to think and the problem moves on without them. Problems get bigger without us. Making space for thinking, soak time as many have taken to call it, really creates a space for the team to ponder and analyze, to think and recall, other important parts of the problem. Soak time is a good defense against normalization of the unusual conditions within which we often put our workers.

Learning is messy and sometimes quite hard to do. The first session should probably not run much longer than an hour or so, and then you should stop the meeting. For the most part, the people will not want to stop; in fact, there will be a lot of pressure to continue on to the solution phase of this process. Stay diligent and end the meeting while you still are in learning mode and there may be more to be said.

Leaving the problem unsolved forces the question to rattle around in the heads of the team members for a while longer. Leaving the problem unsolved also allows the team members to stop the performance of learning to actually take time to learn. Remember, not knowing is a powerful position to inhabit during problem-solving and creating some space for that power to take hold is an excellent thing. In many ways, not knowing is a very important part of problem solving.

What happens during this soak time is the workers will think more about the conditions that are present in and around the work. This additional thinking time will bear valuable fruit for the next time the team gets together. The team will have more to add to the list of conditions when they reform back together as a team. There is substantial potential wisdom in the additional information generated in the afterglow of the first session.

This soak time can be an hour, a day, or a week. Depending on the context of the Learning Team, location, production, or even availability of the people will set the tempo and the schedule for the Learning Team. Whatever time works best for the team will be the best schedule for learning. There are not set rules for how long or how short this break should be.

Don't wait too long. Don't let the fire go out and allow people to move on to other problems that will inevitably arise in their work areas. Everything has its limit and its reasonableness, but know that this break has a purpose … three purposes, in fact … and they all are important to a Learning Teams success:

1. It helps create discipline so the first session stays in learning mode.
2. It allows for additional information, afterthoughts of the team's first meeting.
3. It generates a need for a summary and additional discussion before the next session.

Finally, the soak time period recognizes that the process of learning can be complicated and messy. It is important to stop the group work and give the individual time to think about everything said and identified. It is in these personal review times, these times where we think about what we have done and said, that we start to better understand what and how the event happened.

Learning makes you tired. When you are tired, you don't do your best work. Knowing that this is hard work allows you to build, in short, intense periods of discovery. Don't allow your teams to think they have to have the problem identified and fixed in the first meeting. That will force an early answer and block the deeper learning. Soak time allows for deeper understanding.

## *Phase three summary*

Build time into your Learning Team process for the group to think about what they are trying to accomplish. Allowing time to reflect offers better analysis and problem identification. The soak time also allows a clear delineation between discovery and solution. Soak time can be an hour, a day, or even a week between phase two and phase four of operational learning.

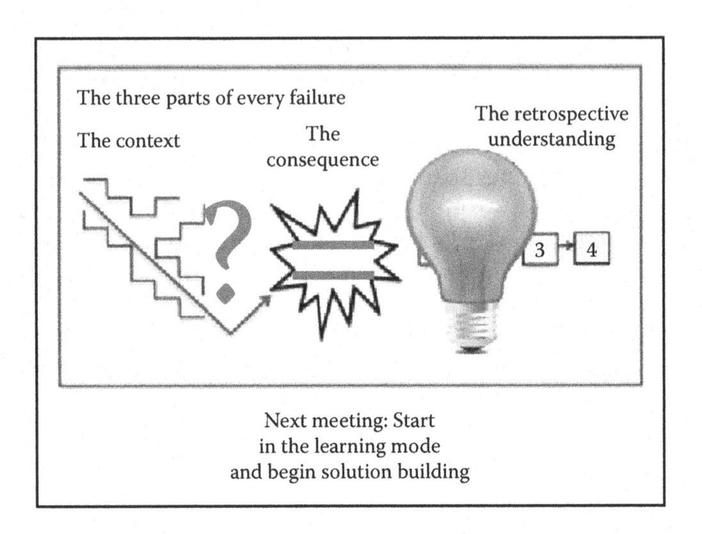


**The Learning Team Phases**

Phase 1 – Determine need for Learning Team
Phase 2 – 1st Session – Learning Mode only
Phase 3 – Provide "Soak Time"
**Phase 4 – 2nd Session – Start in Learning Mode**
Phase 5 – Define current defenses / build new ones
Phase 6 – Tracking actions & criteria for closure
Phase 7 – Communicate to other applicable areas

## *Phase four: The second learning team session*

The team reconvenes and picks up where they left off before soak time, correct? The team then takes all the conditions and issues identified in the first session and create a "path forward" and an accurate explanation of how the event happened or nearly happened. The team does this in a "second session" purposefully scheduled for solution generation.

Perhaps that will happen, but the chances are that a big part of this second meeting will be spent bringing the team back up to speed on what was discussed during the first meeting. It is a good bet there will be the introduction of some new team members for this second session. It is also a safe bet that the team members will have thought of conditions and issues that they did not remember or recall during the first meeting. Your team will want to add to its learning. There will be more things to present and discuss. There will be more conditions and issues added to your flip-chart pages.

The good news is that the additional information is precisely what you want to happen. You want some time to incubate and think about what you missed and thought of later after the first discussion. The process of also reviewing is a process of further defining and prioritizing the conditions that have been listed and discussed. All of these discussions and re-discussions are an essential part of the process. In essence, this second session allows the first session and the soak time to be successful.

This review and reforming process matters. Don't short-change the power of allowing the group to examine and re-socialize the things the team feels necessary and discussion-worthy. This second session builds time for review and re-discussion in to your process. Give it a little time and don't rush it.

This second session allows for some processing time and some definition time to have taken place. We know that after we have had a bit of soak time we have thought more about what we were discussing. This additional time is surprising in that this time allows even more learning to organically transpire. This type of learning is not the group discussion type of learning. This kind of learning is more the contemplative thinking, reviewing type of learning.

This second session provides enough structure to keep the first meeting in learning mode. In reality, what is happening is we are creating a structural division that allows for the necessary discipline (the ability to say, "good idea for a fix – but let's stay in learning mode until the second meeting") to hold the first session in learning mode – not in fixing mode. This becomes evident after your first experience with a Learning Team. Having the second session allows you buy time by saying, "before we answer this question in this session, let's wait for the second session to generate solutions."

Most importantly, it allows the team to collectively review what the team has identified and captured as important during both meetings. This second coming together of the team is where the learning process moves from identification and understanding to knowing and improving your organization's performance. The question becomes how do you guide the team through the second learning session?

## *Review, recap, and capture additional information*

The second session begins by looking at what happened during the first session. Have the team members walk the team through the information gathered during the first session. Allow them to tell you what is important and why. Let them explain what is not important and why and begin the process of taking all the data that was collected during the first session and giving it some meaning and importance.

Session One was really more like a brainstorming session. You captured all the comments and observations without judgment to gather as much information about operations that you could collect. This second session helps you make sense of that data and information.

Have the team tell you which of the information is most important to key our conversation around during the session. What matters about what we know? This session takes what will be pages and pages of information and collectively begins the analysis process. This process happens with the whole team analyzing the work together, learning, judging, and understanding the context of the work.

Ask the team to choose which learnings are most important. Once you feel that the initial findings are complete enough to begin analyzing how work (or the failure or near failure) happens, have the team tell you where the conversations around improvement for this problem should start. This is an important part of the learning process. The team will have the ability to place some prioritization on this list. What should be done first? What should be done at a later time? Which parts of this list provide information for managers and leaders to know and act upon first?

If the idea that big problems are actually just collections of smaller problems, then the process of taking this list of conditions and contexts (really a list of smaller problems) and cutting this information into workable chunks is energizing and exciting for your workers and your organization. I may not be able to fix the warehouse, but I can fix the paint color and visibility on a support beam that has been struck by a forklift several times. More importantly, I don't need to wait for a manager to tell me to fix the paint; I can simply just make the support beam more visible for the worker. Done and done, problem identified and the problem fixed. When workers realize they own the problem, they become extremely empowered in knowing they also must own the solution. The importance of recognizing the problem was not the whole warehouse operation but was the way the support beam was marked illustrates a big problem and makes it a series of smaller problems. More on warehouse beams later ... stay tuned.

The second learning session (or the third or the seventh depending on the issues and team) is that during this session is where we shift our thinking from knowing to doing, from analysis to action. The shift, however, is a thoughtful move towards fixing based upon learning, an informed move towards getting better. We fix not because we have to fix; we fix because we have made the things that need to be fixed so obvious that it would be hard to not improve our processes. All of this process of making the unknown, known is learning. That is what a learning team does, makes the unknown, known.

## *Phase four summary*

Learning happens when the team is not together as well and when the team is together. Allowing time to re-form the group and re-cap what was discussed the last meeting creates an excellent beginning for the second session. Remember, the second session is when the team shifts from discovery to problem solution brainstorming. Allow space for all ideas, without judgment, until the team has exhausted their ideas for making the workplace better. This phase is where thoughts become actions and actions make your workplace better.

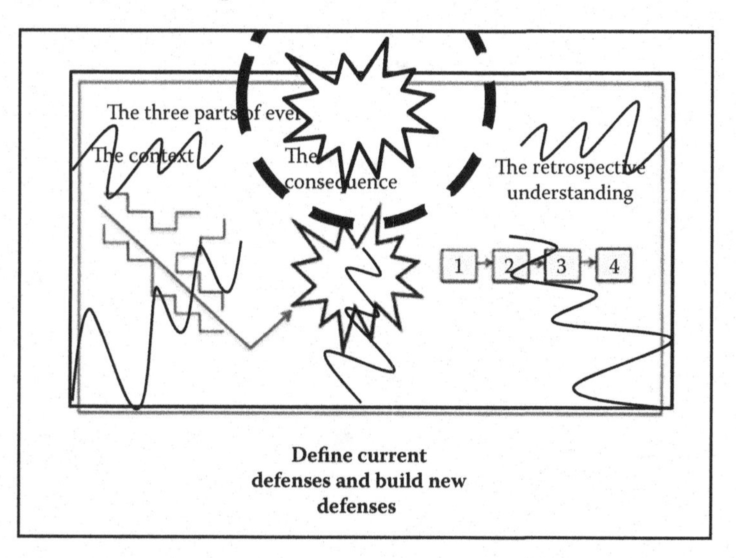


**The Learning Team Phases**

Phase 1 – Determine need for Learning Team
Phase 2 – 1st Session – Learning Mode only
Phase 3 – Provide "Soak Time"
Phase 4 – 2nd Session – Start in Learning Mode
**Phase 5 – Define current defenses / build new ones**
Phase 6 – Tracking actions & criteria for closure
Phase 7 – Communicate to other applicable areas

# chapter fourteen

# Change happens!

## Phase five: Define old and implement new defenses

> Yesterday I was clever, so I wanted to change the world. Today I am wise, so I am changing myself.
>
> **Rumi**

Knowing that deeper discovery will generate a new set of operational information, you will find yourself thinking differently about the current performance reliability programs that exist in your organization. This difference is the point of team-based learning and the reason you include workers in this process. This is a critical part of the process. You have changed the way you are viewing and identifying problems. You are now looking at problems from the workers vantage point, viewing from the sharp end.

The old way we used to identify and solve problems will often serve as unyielding limiters for your thinking. Your current system is the way you have always done things. It is like an old shoe, comfortable, familiar, and a bit stinky. Your organization, for the most part, is almost entirely incentivized to not change the current systems. Changing the system is hard. Changing your system could be expensive (most likely it won't be). Change may mean a lot more work for you and others. Lets face it ... *most* change is hard. Some parts of your organization will not want to change. More importantly, the current system is "how we do the work we do." The idea that you are limiting your ability to learn and improve because of the way the work is currently being done is often scary.

Don't be scared. You are never far away from the best parts of the old system and processes. Change for the sake of change is counterproductive. Not all of your system is bad, much of it runs very stable and very effectively. We must learn what works before we begin to take away what doesn't work.

The reason we always look in parallel at both the current defenses and new defenses is to push the limits of your current understanding of how the work is done. Your teams will learn things that upon identification in retrospect will appear stunningly obvious. That apparent realization that the old system is bad is not a result of your organization being stupid in the past, but a result of looking at the problem in a different way.

For example, I recently helped an organization do a Learning Team for an accident where a fork-truck struck a pedestrian on the factory floor. Unfortunately, the worker was injured by the event. In a fight between soft tissue and fork trucks, the fork trucks usually win. When a Learning Team was gathered and the discovery process began, two significant operational conflicts almost immediately surfaced:

- The central walkway for workers was right down the middle of the warehouse. Every worker who entered or exited the plant did so by walking directly down the central alleyway of the warehouse loading area.
- The three daily delivery trucks were all scheduled to show up at this factory at 9:00 am (the time the event happened). The plant receives three major deliveries per day and over a long enough time that the time became normal all three of these deliveries had drifted to the same exact time during the workday.

During the Learning Team, these two pieces of information seemed incredibly obvious and incredibly error prone when listed on a flip chart. In fact, these two operational conflicts were not unusual; they were very normal. So normal that these two important learnings had not been considered in the attempt to investigate and understand this accident. All the attention of the investigation was placed on workers not paying attention. NO consideration was given to the multiple attention-conflicts that existed in the context of both the warehouse and the accident.

These operational conflicts are so routine that they do not seem to indicate a poor operational working environment. In fact, after years of this operational practice these conditions were not just acceptable, but also normal to these workers. Workers and leaders did not see the additional traffic conflict as bad; these people saw the additional traffic as normal. Stepping out of the process long enough to learn is valuable to the team and to the organization because stepping out allows for learning. Having a different method for seeing work is vital. It is easy for high-failure conditions to exist in a process without much notice. If nothing happens there is nothing to notice. The enemy of safe operations is always stability. The more a bad system does not fail, the more this bad system is thought of as a safe system. When bad things don't happen there is nothing to notice. Not hitting workers with a forklift (traditionally thought of as a good thing) had created an environment where there was an unacceptably conflicted forklift/pedestrian zone. This was a very stable, very bad system. Bad stable systems are incredibly scary. Good stable systems are risky as well, perhaps not as dangerous as a bad system.

A new and different look at the existing protections is a good idea, any time. Doing this same evaluation with workers with the purpose of prediction of failure before actual failure happens is often even more valuable. Knowing where you can add a guard or protection is great prevention work. Remember, our goal is always to reduce the consequence and frequency of events at you facility. Make your current defenses better, but do so by understanding how effective (or ineffective) your current defenses perform now.

It is just as important to know how effective your current defenses are. Sometimes changing your current defenses can make the job *less* safe as opposed to more safe. Not every defense failed if an event happened or nearly happened. In fact, many of your defenses were working incredibly well and will continue to work well. Use care to not throw the good away with the bad. Actively seek places where your system is working well. You workers know where your systems are bad and where your systems are good. After all, they have to use both the good and bad systems every day.

When an accident happens it does not mean all your defenses failed at the same time. In fact, what happened is just enough of your defenses failed, and they failed in such a way, that you had the failure or near failure you had. It is so important to realize that most of the defenses present worked all the way through the failure. Changing your defenses completely is a bad idea. Changing your defenses completely will most often create more failures at your facility.

Let's have a deeper discussion about defenses and safeguards. We know the idea that all an organization must do is identify and place a sufficient amount of defenses within the work environment and all problems will become immediately better is a very attractive, albeit linear, idea. The problem is that defenses and safeguards in a perfect world and without plant variability are seductively unambiguous; they make a complex work environment less complex. Safeguards and Defenses are important, incredibly important. The problem is that defenses bring a completely new set of problems to be addressed by the organization. This is only a caution; defenses make an enormous difference to your success as an organization. Defenses are really attractive to organizations. One can easily fall for the idea that if you put in enough safeguards in a system, the system would no longer be unstable. As attractive as that seems, having enough defenses is never truly possible. Defenses are not bad; there are some hard realities that must be considered when thinking about them.

All defenses have three fundamental truths:

1. Every defense will add a new set of hazards to your organization's processes and work environment. Sometimes those hazards appear later and in different places within the system.

2. All defenses constantly become less effective. Over time, defenses erode and become less and less effective. All systems are running in some type of degraded state.
3. Every defense will eventually by used by smart, efficient, and adaptive workers as some type of production advantage. A worker will use the defense to be less attentive to the hazards to be more efficient at doing work.

Look carefully for the defenses that are working and strengthen those defenses, validate the presence of these defenses, and celebrate the effectiveness of these defenses. Knowing what is working is as important as knowing what failed. In fact knowing what works may be more important in that it allows you to keep the parts of the process that work well while improving the parts of the process that are causing operational pain. You must know both sides of the "timeline" of the failure – what failed and what worked – to actually understand how the event happened. This knowledge is essential.

The risk that you will "toss the baby out with the bathwater" is high post event, in retrospect. It is actually possible to change processes and defenses that generally (and often for many years) work to provide safety for work. We often will throw out all the old defenses in order to make room for the new defenses. That may or may not be the best answer. Knowing what works helps your organization avoid quick removals of defenses by allowing the possibility to recognize the value of the old defenses before these defenses are tossed aside.

## *Micro-experiment your defenses, safeguards, and capacities*

You have heard this before many times, "Don't let great be the enemy of good." Not every solution has to be entirely and immediately diffusible throughout your organization. Frankly, you don't know what corrective actions will work until you try these corrective action ideas on a small, controllable scale. The most important thing you will ever do is to empower your workers to seek problems and create possible solutions, gather information about the solutions, and then improve on these solutions until the solutions get sufficient to be sustainable. Once you have engaged workers, your opportunity to use this same group to experiment these new corrective action ideas is obvious. These folk are perfectly suited to try these ideas and give you quick and clean feedback on the new ideas.

This same idea holds true for safeguard planning and failure capacity within your facility. The best way to test a defense is to micro-experiment the idea in practice. Try ideas that increase safety capacity on a small scale to understand the defenses' validity on a larger scale. Boldly try things, or better yet allow the Learning Team to boldly try solutions to problems. On a small scale, bold ideas are neither scary nor expensive. Bold ideas often provide positive and interesting ideas. Sometimes painting the columns in the warehouse with glitter paint is a good idea, although the concept is difficult to sell for every column in the facility. One or two glitter columns become an interesting experiment. Hundreds of glitter columns are too close to creepy. The crazy thing is that one little experiment with glitter paint led to an exceptionally effective solution for visually identifying parts of the warehouse that are most at risk for vehicle strikes. All of this makes sense when a Learning Team helps the management of the warehouse realize that the paint used to paint the upright columns of the warehouse was the exact same color as the factory stock paint used to paint the forklift. Glitter paint only makes sense when the warehouse is the same color as the forklifts.

Knowing what defenses work and what defenses need to be added to your organizational system is valuable information for leaders to have at their fingertips. These additional safeguards tell managers what to do to make the work safer and better. That type of operational improvement information is why Learning Teams are done. This knowledge is the team's output to your system. It is not about changing the worker. It is about changing the conditions in which the work happens. In many ways our ability to manage conditions greatly outweighs our ability to predict the unpredictable.

Often managers tell me their biggest problem is that they do not know what to fix. I understand that and realize the shortage of information is the problem, not a lack of will or desire. Change is most difficult when you do not know what should be modified. This phase of the Learning Team helps managers know what to change, where to experiment. This knowledge helps improvement ideas become actual improvements. Use the Learning Team to help prioritize which changes should be accomplished first.

There are many ways to determine new safeguards and understand the old defenses present in work. The strength of your organization's defenses is in constant flux, based upon contextual factors surrounding the work. Remember that because a defense failed once, does not mean the defense was always weak. Defenses drift towards failure. No defense is ever permanent, and defenses are normalized and weakened all the time. Defenses change and adapt, just like work changes and adapts. Don't mistake your system (or your system's defenses) as permanent and unchanging.

*Phase five summary*

Learning Teams help understand which defenses worked and which defenses need more attention and time from leadership. Knowing what works (and why these defenses worked) is vital to knowing what to fix. Full-scale replacement of defenses will upset the current organizational structure and may make the work more failure prone. Leaders always have the ability to try out new ideas on a small scale to learn how these defenses can be more efficient.

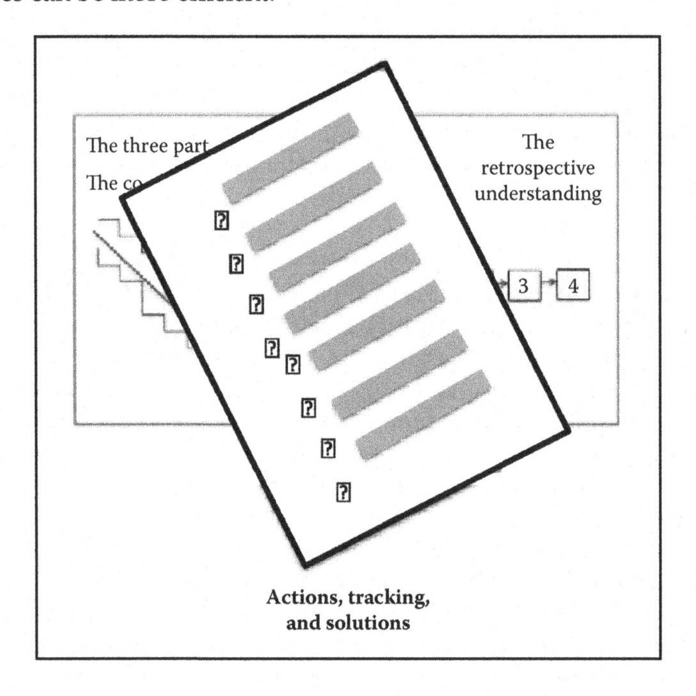


**The Learning Team Phases**

Phase 1 – Determine need for Learning Team
Phase 2 – 1st Session – Learning Mode only
Phase 3 – Provide "Soak Time"
Phase 4 – 2nd Session – Start in Learning Mode
Phase 5 – Define current defenses / build new ones
**Phase 6 – Tracking actions & criteria for closure**
Phase 7 – Communicate to other applicable areas

## *Phase six: Track actions and criteria for closure*

We must stop measuring safety effectiveness
by counting the number of people, we hurt.

Learning Teams not only help you identify and solve operational problems, they also are effective in giving you another prioritization data set. Learning Teams help you manage your prevention resources better by allowing the worker to have direct feedback on what you should be fixing first. Members of a Learning Team know which fixes are easiest, which fixes are most effective, and which fixes can wait for budgets, workers, additional resources, and time.

The work product of a Learning Team is information. Not all of the information that the team provides to you is equally important, equally significant to your operations. Knowing which improvements will have the most substantial impact is a mutual decision based on the input from the Learning Team (the process users) and leadership (the process owners). Having both voices come to the table actually gives better prioritization outcomes for your operations. In a way, this process helps you make some very difficult, resource-constrained decisions about your operations.

The problem then becomes what is most important. What should we fix today and what should we fix tomorrow? The team can and will answer that question for you. One significant benefit of a Learning Team is the almost instant correction of latent negative conditions that exist in your operations. The bad stuff gets noticed and taken care of almost immediately. Much of what the team identifies as a problem this same team is now empowered to go out and fix. In most cases the team will immediately fix the immediately fixable conditions.

There are basically three reasons workers don't fix problems on the production floor as soon as they discover problems in your operations. Oddly enough, none of these reasons involves being lazy or stupid. The first reason workers don't fix problems is the worker does not think they are supposed to fix the problem; somehow fixing whatever needs fixing is not their job. The second reason workers don't fix problems is the belief that if the worker fixes the problem, they will get punished for their efforts. This second reason must be some type of further definition of the first reason. The third reason workers don't fix problems is the belief that they do not know how or are not empowered to get the necessary resources that would be used to fix the problem.

1. It is not my job to fix it.
2. I will get into trouble if I do fix it.
3. I don't have the authority to fix it.

Learning Teams help to dispel all three of these reasons workers don't fix problems in your operations when they see the problem. You can see how this level of engagement and empowerment, coupled with the ability to do small experiments on the production floor, does allow workers to identify and fix problems that they live with all the time.

Yet, these ideas don't give managers and leaders much to measure. You can count the number of Learning Teams that you have had in your facility. You can count the number of problems identified and fixed. The problem is these metrics are neither relevant nor predictive. The better you get at safety, the fewer events and problems you have to measure. In short, measuring what does not happen is difficult and unsatisfying. We have a problem and that problem is if we are good at learning and improving, we don't have anything to measure using the traditional safety metrics that exist in our organizations.

Let the discussion begin. Here is the most asked question of any organization that is currently trying to move their safety and reliability needle towards the new view: "What metrics do you use in your organization to tell how you have improved?" What should we measure? My leadership is currently demanding metrics? How do you know what you are doing if you aren't measuring anything?

How to answer these questions is related to how well you understand complex operations. Some things we measure are not that important but are very easy to measure. Some things we struggle to measure, and so we ignore the metrics about that topic. Everything can be measured, or at least it would seem that you could measure anything that seems relevant to your organization. The problem is some of the improvements you are making are not very easy to measure. That is a dilemma!

Complex systems do not lend themselves to specific metrics, measurements of specific milestones in time or in a process. A river is incredibly difficult to measure as a river; you can measure the amount of water that passes a gauging station; the height of a river is measurable at a certain time, for only that time. You can measure how wide or deep a river is – at a fixed point on the river itself – that is also measurable. You can even count of the number of fish that pass through a certain area of the river – that is knowable as well. The problem with all of these measurements is that they only represent the river at a certain time and certain place. They are metrics to be sure; they simply are not metrics about the river as a … river, living, flowing, running, vibrant, productive river. A river is always a complex biological system, constantly adapting.

In many ways, the example of measuring the river is our exact challenge. How do we measure a group of people who are forming around a problem? The answer goes right back to complexity. We must think of a Learning Team as a complex learning capability and look for progress in that way. In short, you don't measure a river with a metric; you can't actually measure a Learning Team at a fixed point. I recommend that you measure movement for the Learning Team. Is the problem getting better? Did the problem get worse? Is the team moving forward? Is the team functioning? Are your processes getting better or getting worse? Does the team feel as if they are learning? Those metrics are much softer, but also much more accurate for this type of work. Determining when the team is a success is much easier if you establish criteria for improvement as opposed to a set target for success. Most teams will be successful although it may not feel like the learning the team is discovering is a success at the moment.

## *Learning happens in many ways*

I was working with a Learning Team that was made up of line workers at a rather complex manufacturing organization. The organization was not too complicated, no more complex than most organizations. The work these particular workers performed every day involved multiple hazards: heat, pressure, mechanization, flame. The only hazard that seemed missing were either bees with their tails dipped in poison or sharks with lazers on their heads. The problem this learning team was seriously working to solve involved a piece of "gently used and loved" equipment that had been inherited from another facility. This machine was promised to work without any attention, no maintenance, little setup, and this incredible machine would save loads of time and free-up three workers to go to other places in the facility. If this machine had worked as promised, work would get easier and much more profitable. This machine was going to be a huge improvement for this facility.

Needless to say the machine did not perform as promised, and this machine did not have any engineering documents that traveled with it. The machine was substantially difficult to fine-tune by experienced operators in the best possible conditions. Imagine how a group of supportive, albeit inexperienced with this particular piece of manufacturing equipment, workers worked repeatedly in order for the machine to function somewhat closely as promised and needed. If you imagined this problem in classic used car sales terms, these guys had been promised a Rolls Royce and been given a lemon.

In my mind, I framed this problem as if a person was buying a boat; the seller made many promises that the boat they were selling (not the boat they were sold) simply could never, even on the best of days, deliver. "Sure it runs like a dream and never leaks!" "You will love this boat." ("Only ever used by an old lady who drove it to church on Sunday.") This machine was not operator-free; it needed tremendous love and care. It was learned, as the team was deep in their discovery phase, that this machine needed constant attention, love, and care in the former factory in which it was used. It will require the same or more attention in the new home for the machine. Only this time, the experienced operators were not present. The new operators will have to learn how to make the machine work while at the same time, building their own experience-base for the plant. This was the problem that this organization had to solve. This was a big problem.

The plant manager put a Learning Team on this problem and the team began to identify the conditions that were causing the most significant problems. The team took its time to dig deeply in to this problem and discovered many problems. Eventually, the team added a young engineer. The engineer brought in some folk who could reprogram the computers that operated the machines. Soon this new machine was not performing as poorly as when it was newly installed. The machine was not a lot better, but the machine was better.

About a month later, we did an operational learning workshop at this plant and the plant manager spoke of this Learning Team, their very first attempt at team learning, as a failure. He still thought the idea was good, but this problem was still not solved. The team had failed in his mind. He assumed that this problem was too big and too complicated for workers to handle without the power of management being present in the room. The plant manager was so certain the first attempt was such a failure, he offered this problem, this machine, and this Learning Team up as the workshop example problem in order to take this problem toward a much needed solution.

When the workshop took on the installation of this machine in the plant as the workshop example, the workshop Learning Team did exactly what any Learning Team would do: we started looking at the conditions present that are creating the failure of this piece of equipment. Interestingly enough, when we started going over the conditions with the workshop team almost all the issues that were identified already had a correction path in action. It was clear to the workshop group that the initial team of workers on the first Learning Team had not failed at learning. Quite the opposite ... the team had taken this problem to an entirely different level.

The second cut on this problem was much different than the first cut on this problem. The initial Learning Team had learned and fixed a whole lot of the issues present in this problem. The team was developing a capacity to understand this issue. The team had yet to realize they had this capacity, and therefore, lacked a bit in the confidence to make the problem better. The plant manager was too close to this problem to even realize that progress had been made, was being made. In his mind this problem had two states: fixed and not fixed. The machine was not fixed; the plant was not fixed yet.

When the initial team, the Learning Team, was brought into the workshop in order to combine and leverage the old learning with the new learning, it became abundantly clear that much had been learned, much was known, and much had happened. One of the mechanics on this machine spoke in glowing terms about the young engineer that listened to the operators, took their comments, and made advancements in the machines operations. Not realizing the young operator was in the room listening, he was really open and honest about how valuable this exercise had been. It suddenly became clear that the initial Learning Team was not a failure but was an amazing success. The plant manager was both surprised and pleased. Funny, that a half hour before this meeting, the plant manager was very frustrated by this problem.

There is much for us to learn from this example. Progress does not necessarily have to meet management's tests for success, especially if management is looking at a "point in time" success metric. Development happens in many different ways and at many different speeds. That young engineer was so proud, I am certain he will use workers in his engineering learning for the rest of his career.

The plant manager asked to speak to the group towards the end of this particular workshop. What he told the group was that learning is not what he expected learning to be. He had mistakenly thought that learning was done to produce answers. In fact, learning is done to generate wisdom and knowledge. Knowledge and wisdom are then used to create solutions to problems. Operational learning happens because the group was brought together, not by the answers the group either found or failed to find. The process brought the right people around a difficult problem and started the process of improvement.

I could not have been prouder, but my pride hardly mattered. The pride in this workshop belonged to the plant manager, the young engineer, and the Learning Team that had started the process of improving a very delicate piece of equipment that was not working well. Funny, now the machine works better than it did in its original plant with its original workers. Now the machine is producing beyond the promised production numbers. Seems the new machine operations computer code coupled with a new cleaning platform has this process producing material beyond specifications.

## *The power of small experimentation*

The best way to know if an idea is a good idea is to try the idea in a real-life application. We often don't try small solutions because we have been taught to wait for the grand solution to our problem. For some reason, we sometime forget the power of an operational prototype. Our organizations believe that the planning process is the same as the testing process. We do this all the time. I work with organizations that just don't test their ideas on a small scale because the organizational people are too busy trying to fix the problem for the entire facility. You cannot change everything at once. "Big Bang" rollouts usually fail. Learning is your friend. Testing always leads to more information and more information means better learning.

Small experiments allow you to quickly collect data about your solutions and analyze that same data very quickly. Trying out new defenses, safeguards, and barriers allow for small samples of what may work and what definitely will not work. Trying out ideas allows the organization to carefully begin the process of change.

Small-scale experiments offer another advantage for learning. A small-scale test gives you room for recovery if the test fails. Even if it is the most stupid idea in the known world, you have not invested a lot of time, energy, or effort for the stupid idea to cause much damage to your organization or to the member of the Learning Team. In many ways, small-scale testing is a safer way to see if your team's ideas have merit and the potential to scale to a larger solution.

Perhaps the paramount advantage to small scale testing is the power to retract ideas that are not working. It must be safe to fail in your organization. Solutions need space to be successful and space to be bad. Knowing that the solution is a small experiment allows the team to pull back on the solution idea without leaving a giant trail of harm. You can retract without fear of long-term harm or embarrassment.

### *Success has many faces*

The most important judge of when learning has happened is the Learning Team. This team is having the discovery and so this same team is probably the best judge of when the discovery journey is complete. Completion may not appear the same for all levels of the organization. Perhaps a better way to understand the idea of completion is to not think of being complete and getting to a state of "no problems left," but more as a motion towards a solution. We know from our discussions that we don't see complexity as a fixed point in the organization's processes, but as action towards the creation of value. Did the team make the problem more understandable and more transparent? Is it now easier to explain what is happening to better understand how to make the work better and safer and more reliable?

A big part of the journey your organization is taking is the redefinition of what success looks like to your site. We must teach ourselves how to improve and improvement is sometimes slow and incremental. Improvement must also be deliberate. Your organization improves on purpose, not by chance or accident. Knowing that success comes in small, incremental steps is essential to understanding when your team has accomplished what the team has set out to achieve.

You must also be open to the fact that what you learn from these Learning Teams may not be what you set out to learn. Most of the time, when a team takes on a problem, the solutions reach deeper and broader then the problem itself. Having a different answer is also a success. All improvement is improvement.

## *Tracking actions*

Out of all this movement will come some actions, some fixes for your problems. Not all of these actions will have the same importance. Some actions will become crucial to the team. Some actions will become essential to leadership. Some actions will be essential to both the team and the leadership. Allow the team to help prioritize which activities are most important. Knowing that, this idea of importance is a perception of individual priority and is based on where you are in the organization and what is important to you.

As the team identifies conditions present in the Learning Team's target, the team will also begin to decide what can be done with these circumstances. Some of the problems are extremely easy to fix and other conditions can be costly and time consuming to remove or mitigate. That is normal and quite understandable. This is where your team can help you as a leader, understand what actions should be taken to create improvement.

Be careful to not overload your system with gigantic lists of improvements. Make sure the team does some pretty aggressive prioritization in order to create a list of a select few improvements that will have improvement impact. Immediate findings, findings that if not fixed could cause serious harm should be fixed immediately. Improvement findings will benefit from a careful discussion of how a select few changes will be easier, faster, and less disruptive to the overall facility.

We typically ask the team to use the list of "fixes" the team has identified as a way to discuss what actions should be taken first, second, and third. What are the easiest and best immediate actions? What will take more time and a bit of resource? What longer-term solutions are going to be much more complicated and expensive to actually improve? Think of these as "one star," "two stars," and "three-star" solutions.

The team has the knowledge and the depth to know the problems that can be fixed immediately (in my experience most of these immediate-fix problems will have already been fixed by the second meeting!). Allow these issues to be fixed. Enable the team to try-storm some solutions and see how they can make these problems better. If a new parking area is discussed for the fork trucks – take a roll of tape and restripe the floor with the new method. If this new method works, add paint and permanence later. Any initial early wins should be allowed to win. One star solutions are the low-hanging fruit of performance improvement; you probably could have (and often have) fixed these problems, but, in this case, allow the team to make and take credit for these changes. Remember, progress happens in small steps – improvement is not instant. We are building confidence and capacity in our workforce to identify problems and to fix problems. There is incredible power in positive reinforcement.

The next level of action may take more time or more resources to be used as a defense at your site. These actions are marked with two stars. Two-star solutions will take preparation and time, perhaps a bit more resource use as well. The two-star solutions are the solutions that management usually goes to in traditional organizations, organizations that are not quite ready for the new view. Two-star solutions are necessary, but failure to immediately solve two-star solutions should not stop progress on the one-star solutions.

Solutions for prioritized problems move forward many facets at a time (like a blob of ooze taking over a city in a science fiction movie, or worse yet, in real life). Unfortunately, problem solution does not move in a predictable, linear fashion. We fix problems as we identify the need to fix the problem, not as if we are using some type of strategic plan. Learning is messy and fixing is messy, too.

Workers are especially good at knowing what to do "now." Workers often lack the knowledge and ability to know which action makes the most sense for more long-term solutions. One of the reasons Learning Teams are created is to give leadership the best operational information possible. Information about the "now" combined with information for the "long-term" give leadership a better, more holistic understanding of what to do at the site to improve performance.

Learning Teams should fix what they could fix immediately and use what they have learned to help leadership plan further, more permanent improvements for the organization. The best test to use to tell if a Learning Team is finished learning is the test that asks, "has the team learned enough about the problem that the team has a better understanding of what is wrong and what to do?"

### *Phase six summary*

We know that learning is sometimes messy. Learning success is dependent upon the perception of the team that is learning. Learning is often small bits of knowledge making up larger discoveries. Small-scale testing of these ideas is an effective way to test your ideas for effectiveness and scalability.

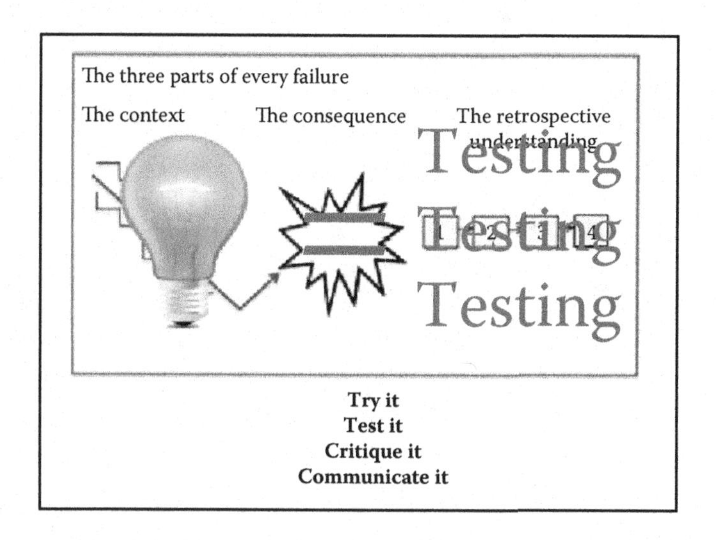


**The Learning Team Phases**

Phase 1 – Determine need for Learning Team
Phase 2 – 1st Session – Learning Mode only
Phase 3 – Provide "Soak Time"
Phase 4 – 2nd Session – Start in Learning Mode
Phase 5 – Define current defenses / build new ones
Phase 6 – Tracking actions & criteria for closure
**Phase 7 – Communicate to other applicable areas**

# chapter fifteen

# Shout from the rooftop

> Good news is rare these days, and every glittering ounce of it should be cherished and hoarded and worshipped and fondled like a priceless diamond.
>
> **Hunter S. Thompson**

## Phase seven: Communicate to other applicable areas

We tend to only tell stories of safety failures in our organizations and we seem to take the success for granted. The noteworthy thing about telling workers scary stories of what could happen to them is that these stories don't scare workers – it did not happen to them. The workers feel like the chances of this failure happening belong to the poor worker that had the event. There is a 100% chance that the worker who had the event will have the event. There is also a 100% chance that the worker who did not have the event will not have the event. We have to tell stories of safety successes. We must!

Teams need to feel successful and valued. Teams need to feel that they have learned something that matters and will make a difference. Those same teams have committed their time and their attention towards improving your organization's work environment. The team has skin in the game and they care. Your job is to make sure that the team feels the importance of what they have done. Remember the feeling "that my opinion matters, helps me build confidence."

Several times during this book, I have mentioned the words "confidence and capacity." In fact, the idea that we build confidence and capacity are words that I find myself saying repeatedly. What engaged learning does is build the confidence that the workers can discover and identify the problems that exist in your work processes and systems. Once workers know that their ideas and observations matter, are heard, and are unique and accurate, the worker will take his or her responsibility to look for upset conditions much more seriously.

The next move is to give these same workers the capacity to understand and fix the problems identified. These workers don't have to fix the problem for the entire facility, but indeed these workers are empowered to fix the problems that most influence their ability to safely accomplish work. This combination is mighty powerful. This combination is a very powerful method to use to change the safety culture of any facility. It works. It works exceptionally well. It only requires that you as a manager lets go of a bit of the perceived control that you have in learning about your work processes and systems.

## *Constantly search for "extent of conditions"*

A vital part of learning and acting is determining, "where other failures can happen – knowing what you now know?" One of the outcomes of knowing more is being able to predict the "like conditions" that may exist within our organization. Where does the learning we have discovered from this event apply to other like places within our workplace? Find the other areas of your organization where this learning is applicable. There will be many.

Knowing where else an event can happen, the extent of conditions may or may not be a part of the charter for the Learning Team. If you don't make it the work of the Learning Team to address where else an event can happen, you may be missing a chance to engage some very smart people in this expansion of value process.

Identifying other places can seem overwhelming at first; don't let it scare the team. Doing an extent of condition review can seem significant and complicated; don't let the fact that you can't and won't find *every place* possible stop you from looking for *other* places where this event can happen. In this case almost any information is better than no information.

Build fluency around the language of the extent of conditions into the vocabulary of your organization. Help your team feel empowered enough to think about other areas of the facility where these same *conditions* may or do exist. The secret is in the identification of conditions over consequences (or potential consequences).

We have been taught our entire career to manage consequence or potential consequence. The very idea that we manage consequence biases our thinking about where else an event could happen. In many ways, the hunt for like failures, you could call these hazards, causes us to limit our thinking to the identification of where bad things can happen. Bad things can happen anywhere, that information is very valuable and a bit overwhelming. Many times, I have been involved in discussions with leaders of high-risk operations who say, "Everything on this site can kill you." If everything is dangerous, then nothing is dangerous and everything becomes a hazard to identify.

What we want to identify is where else the conditions that lead to bad things happening exist. This process of understanding and defining conditions is much simpler and much better (and much more measurable) than predicting the future of your organization based upon the many ways a worker can die. Conditions, if identified in some order or amount, are easy to see. Identifying conditions leads to looking for real things, the parts of a failure that must be present ... that can lead to a failure.

To the extent that these same conditions discovered in your learning may exist in other parts of your operation is the very reason we do Learning Teams. Learning Teams are an efficient way to understand a current problem. Learning Teams are even a better way to predict and render harmless a problem that has not happened yet. I can't understand why, in some industries, organizations don't know that the learning is not finished until the extent of conditions review has taken place. However, many organizations don't do this vital part of the learning activity. The extent of condition review is often as simple as asking the Learning Team members this question, "Where else do we have these same conditions? Where else should we apply this learning in our organization?"

## *Phase seven summary*

The entire reason any investigation or Learning Team exists is to create more learning to facilitate better improvement. All investigations happen to learn and improve. Knowing when to stop learning is as important as knowing when it is time to learn. Not everything that happens needs or requires deep operational learning. Knowing where else this knowledge can be used is a shared responsibility between the team members and the operational leaders.

# chapter sixteen

# A learning team case study

**Bob Edwards**

## Who buried Bill?

A team of county road workers was repairing a washed out culvert. They had one of their guys, who we will call Bill, down in the trench doing some hand shoveling work when part of the trench wall gave way and trapped his legs. His fellow workers quickly dug him out and had him taken to the hospital to be checked out. Fortunately, there were no significant injuries. However, everyone realized that this could have been an atrocious event. Many trench engulfment accidents cause major injuries and often times fatalities. The county leadership team wanted answers and they needed to know who was to blame for this event. In the discussion about the event, a good friend of mine who happens to be a councilman, thought it might be a good idea to have the county officials sit through a human performance presentation. Perhaps they would see that there were system weaknesses that needed to be corrected instead of looking for someone to blame and punish. When my friend approached his fellow county officials and presented this plan, they were willing to try it. I think most of them thought they already knew what to do; they needed to find out who screwed up and discipline them so that this sort of thing won't happen again.

When I showed up for the briefing, I began with the usual conversation that Todd and I lead a team of managers through to see if they are interested in moving past the blame model. I found that as I presented the information, it was pretty well received by most everyone in the room. I like to use real-life examples as I teach the principles of human performance so that it makes sense in a genuine and applicable way. As I discussed the need for operational learning around an event like this, I shared with the managers the importance of open and honest communications with those who were close to the event. I shared the need to listen, learn, and not go in with preconceived ideas about what caused the event. I described the basic steps we use with operational Learning Teams and stressed that a critical part of the discussion is centered on building trust with the Learning Team.

The example, I used to help illustrate the importance of trust came from another Learning Team that I led earlier at a site that had experienced an equipment fire. I told about a discussion held with three operators, a local area supervisor, the safety manager, and myself. In this fire story, I was able to show how trust was built over the period of about one hour. As I had facilitated the fire event Learning Team, two of the operators in the discussion were telling me everything they could about all of the conditions around the incident. As they explained all of the complexity and issues that led to the fire, I only continued to ask questions and write their responses on flipcharts. It was evident that no one came to work and said "Hey, let's set the equipment on fire!" If they did that, it's a different story altogether. It's time to call Homeland Security! In fact, what had happened was the first shift operator tried to run production and the equipment kept double loading plastic parts. Eventually, one of the extra pieces fell down on the heater coil located below the part transfer table. The heater coil was used to pre-heat the plastic before forming. During the attempt to remove the melting plastic, the material caught fire and the operator had to use the fire extinguishing system built into the machine to try to put the fire out. Another problem arose as they hit the extinguisher button; the extinguisher nozzles were not set up for extinguishing a fire with the part transfer table in the location it was in when the fire occurred. When the operator hit the extinguisher button, the extinguishers unloaded their entire contents right below the part transfer table and heater coil and onto the floor. Not good! The operator had to run and get a large hand held extinguisher and then climb part way up on the side of the machine to spray the fire and finally put it out. As the operator explained all of the issues, the second operator shared with me several issues that he had noticed with the equipment as well and that he believed contributed to the fire. I saw the third shift operator sat there quietly and didn't say anything for a long time. I didn't pry; I just kept asking the other two operators questions and continued to write down their comments. Towards the end of the one-hour learning session, the third shift operator raised his hand and sort of sheepishly said, "I wasn't going to say anything. But, I have actually had two fires before; I just didn't say anything about them." Wow! Now that's interesting. I realized that how I responded to this statement was incredibly important. If I said something like "What? You didn't report them? Why? What is wrong with you?" the learning session would be over and everyone would shut down. Instead, I just asked him to tell me more and I turned around to the flip chart and kept writing. When he realized that I was not there to try and blame someone, he became willing to help. In fact, he was able to help us understand even more about the system weaknesses and equipment issues since he had dealt with two fires.

As I finished telling the "trust" story, I noticed that one of the county officials had her hand up. It was the Human Resources Manager. When I inquired what was on her mind, she told me, "I like a lot of what you are saying here today. But you are telling me that you didn't write that third shift operator up for not reporting?" I replied, "No, we didn't. I was just glad that he was opening up and sharing his stories with me and the team." She couldn't let it go, though. She continued to disagree. She told me about her "open door policy" and that her county workers came to her with problems. She also stated that if the third shift worker had told about his fires that we might have avoided the third fire. My response was simple: "The only reason the third shift worker was even willing to say anything was because we built a "trust environment". He saw that we were more interested in understanding the conditions that caused the fires than worried about who was to blame." She kept going back to the fact that we should have disciplined him and that if the site had an effective "open door policy" he would have reported the previous fires. I realized that the best thing I could do for this HR manager was to invite her to the Learning Team session that afternoon and that's exactly what I did.

After lunch, they brought in the team of road workers to have a "discussion" about the trench engulfment incident with Bill. They were all wearing their bright orange road worker shirts and looked as though they had just come in from the field. For the most part, I could tell that they were not that excited about being there. Most likely, they feared that management had brought in the "outsider" to find the underlying cause of the investigation and determine who was to blame. Almost immediately I tried to put them a little more at ease by telling them that I was just a safety guy from a local company. I was asked to come in to lead an open conversation about the trench event. I told them that our goal was to see if there were weaknesses in the system that resulted in the event. I still think most of them were a bit skeptical, but at least they knew I wasn't from the "Trench Engulfment Accident Investigation Team!"

I started with a brief introduction to the basic concepts of human performance Learning Teams. I explained that I wanted to take some time to understand their world and learn about the type of work they do. The first question I asked them didn't have anything to do with Bill's incident. I merely asked the room full of road workers to tell me the type of work they do on a daily basis. I could see them looking around at each other with that look in their eyes as if to say, "Aren't we going to talk about burying ole Bill?" So that the reader understands, whenever I lead Learning Teams, I never start with a discussion about the event. I find that a better approach is to start back in the process and move the conversation forwards towards the event. It provides a much better conversation. It also helps the team get away from the emotions that often accompany the event itself. If someone was hurt or may have been punished, there is just naturally emotion there. One of our goals with Learning Teams is to move away from the emotional reaction and instead attempt to learn and respond.

To help clarify, I will take a minute to explain my journey in the world of investigations. In the past, as a safety leader, when investigating an accident or injury I would start at the event and ask "Why?" five times looking for the "root cause." This approach seemed to satisfy my basic need to understand and explain why the event occurred and then gave me the ability to fix something. As a safety leader and experienced engineer, it seemed like a concise method and provided me with an explainable linear path that led to the failure. Fix the root cause, and the problem was solved. The problem with my thinking was multifaceted. First, I have found that I was so focused on finding a "root cause" that I was not looking at all of the conditions that led to the event. As I began to change my approach, I noticed that if I focused more on learning and discovery instead of fixing. I ended up with a more comprehensive story of "how" the event occurred, instead of just "why." It has also become quite evident to me that failure is not as linear as I thought. The 5-whys approach tended to build a linear path of what appeared to be "cause and effect," where I think it actually was more of an "effect and cause" picture of the event. I would see the effect and then sort of create or determine the cause. The more Learning Teams I led, the more I realized that failure is actually incredibly complicated in most situations.

It now seems to me that there is no real path to failure or "chain of events" as we used to talk about in the Army. I have reached a point now where I don't think there is a "root cause" or even "root causes." It seems to me that many of the things that lead to failure are simply normal variability that eventually aline in such a fashion with other conditions and hazards that lead to something bad happening. In fact, the search for root cause had often led me to believe I had fixed the problem when actually I only fixed one piece of the failure; perhaps it was only the triggering event. In reality, there were numerous other conditions that coupled to bring about the bad event. So it's the Learning Team's job to define the "how", "why" and "what did it take" for the event to occur instead of just the "why."

Returning to the story, as the road workers began telling me the type of work they do, I turned to the flipchart and began to write. They explained that they use backhoes, dozers, bobcats and dump trucks. They told me about their work with fixing roads, culverts, ditches, and bridges. They told me about the long hours and sometimes harsh work conditions in which they have to perform their jobs. Then I asked them an interesting question. I asked them to tell me about some of the near misses they had seen in the past. You want to hear some interesting stories? Talk to some road workers about near misses. As they began to tell me the stories of their world, I noticed the HR lady was actually paying attention. She heard stories about issues and situations that she had never heard before. She became super interested in what the guys were saying and I believe the light bulb came on for her. She realized that she would probably never have heard any of these stories if it were not for this conversation we were having now. I think she saw that her "open door" policy was not going to bring tough ole road workers to her office. As a matter of fact, if you think about it, I can't imagine any of those guys ever experiencing some near miss out on the work site and him saying to his fellow workers, "Hey, that driver in the car just flipped me off. I'm going to HR about this!" If that ever happened, I would imagine the other road workers saying something like, "Hey you big baby, shut-up and get back on your backhoe. Nobody in HR wants to hear you whine!"

So next, we began to talk about the days leading up to the event. It had been raining a lot. The ground was super saturated. The culvert washed out. They were under pressure to get the road re-opened. Civilians were unhappy it was taking so long. They had no trench boxes or training on how to use them even if they did. They were doing everything they could to get the culvert replaced and while they were installing the new one, there was a need to level out some of the gravel in the trench. Bill grabbed a shovel and jumped down in there and began leveling it out; without warning, the sides of the trench gave way and collapsed in on Bill's legs. Fortunately, it only buried him about thigh high and the guys were able to quickly dig him out of the trench. He was taken to the hospital and treated for a sprained ankle and that is all. Therefore, this event was one that could have been much, much worse.

I then took the team through a conversation about what they thought would make things better. This team of road workers did some homework as well and determined what sort of equipment and training they needed to conduct this type of work safely in the future. If the team solves the problem, they own the solution and are much more likely to keep the solution in place. If the management solves the problem, it is often seen as a mandate from someone who doesn't really "get it" when it comes to getting work done. I also notice that the Learning Team usually comes up with more thorough and meaningful solutions since they are close to the work and the event and they don't want it to happen again. This empowerment gives the team more confidence, which in turn builds more capacity to solve tougher problems in the future.

An interesting side note is that the county officials want to have additional training on human performance and operational learning. All of the departments that is, except one. There was one department from the county that was not interested in learning any more about human performance. Their manager stated that he was afraid of losing accountability with his employees. Funny that they tend to treat their employees with the blame and discipline model and that they are not interested in improving or changing the way they treat them. That department was none other than the sheriff's department. I guess they are so used to dealing with people who intend to do bad things that they are more comfortable with the disciplinary model for the citizens of the county and their employees. Oh well, we can't win them all over!

# *chapter seventeen*

# *Conclusion*

## *This book ends and your work begins*

This is from my email:

> Question:
> Learning Teams are the best safety engagement tool I have ever used.
> My secret is that a Learning Team is faster, easier, and better than anything else I have ever tried to engage workers. Why is this working?
>
> Answer:
> What you are building when you focus on engaging people at all levels in your organization is both confidence and capacity. You are building the confidence to know the worker's opinions matter and the ability for them to do something with those opinions.

### *Learning and improving*

Why does your organization do investigations?

I have taken to asking this exact question to organizational leaders and I am constantly surprised, frankly shocked, with the answer that these leaders give to this question. Most leaders seem quite surprised by this question. They tend to think a moment and then tell me a rehearsed-sounding answer. Most leaders tell me that investigations are done to stop repeat occurrences of an event.

That answer gives me pause. In fact that answer seems wrong.

I never want a repeat occurrence of an event; no event should ever happen twice. The problem is that no event ever happens twice. Context for events is always different. We never have the same event happen. Every event is singularly unique in the conditions that were present to create the context for the unwanted outcome. Therefore trying to stop the same event from happening again is not helping us get better. In fact, that type of thinking probably moves us towards ineffective corrective actions and event explanations.

The only answer to why any organization does investigations is, "to learn and improve." We do operational learning of all types to get better, to learn, and to improve, or at least we should. Investigations help us get better. Investigations provide explanations for how failure happens in our work environments. Investigations allow our organization to learn and improve. The fact that good leaders have decided that we investigate "to never have the same event happen twice" leads our organizations to very limited and incomplete learning. Teach your leadership that every operational critique and investigation we do is to provide more and deeper understanding of how our organization works both in times of failure, but more importantly in the times when successful work is being done.

The challenge is to get better at learning so we can get better at improving our organizations. We should not study what happened, but rather what is happening. In order to shift from avoiding repeat occurrences we have to change the questions we ask and the people to whom we ask these questions. We must involve the people who know, who do the work, and who know what is happening in our organizations. Once we have a better understanding of what is happening, we can then engage the same population in helping us improve. That is the entire function of a learning team.

## *The old is new again*

Engaging workers in problem identification and problem solution is not new or unique. Organizations have done engaged learning since time began; yet, in the world of safety and performance improvement, our processes have fallen prey to the idea that "management must know better." This book has been a discussion of how wrong that idea must be. Managers have big picture knowledge of work with little-applied experience. Managers and planners know how work is meant to be done – the one right way to do a task. Workers don't have the big picture view, yet they are experts in how work actually happens. This separation in thinking has changed the way we identify problems and how we solve problems. This separation is not a good one, nor has this separation served us well in our constant quest to learn and improve.

We have the wrong people, the wrong part of the organization, telling us what needs to be done to understand how failure happens. Because we are using the wrong people, it goes without saying that we are getting the wrong information about the failure. That puts the leaders in a position where they are making the best decisions they can possibly make with the worst information they could possibly use to make those decisions. That is our problem. That is also our opportunity. Holistically understanding these problem/opportunity data sets is a fast and effective way to change the future of your organization.

When we give our leaders good operational information about success and failure, our leaders make better decisions about our path forward. In a sense we are improving the in-puts, the information that is being fed in to our performance improvement system, and in so doing, improving the output of our performance improvement system. The result is improved performance and learning. Not a bad day's work.

Your work begins by building the capacity for your organization to learn and improve. The only way to build this type of capacity is to begin building this type of capacity. The more you learn the better you get at learning. Long journeys begin with one small step, so too does learning begin by beginning to learn. You will have to have a "first attempt" at worker-centered problem identification and problem solution. The first one will build the confidence to have the second one, the second one builds more confidence for the third one, and soon you will have done hundreds of learning events. But, everything depends on having the foresight and the fortitude to have the first learning. I promise, the more of these "learnings" you do the better you and your organization will become at learning.

## *Why learning teams?*

The short answer is without engaging your workers in the process of problem identification and discovery and the creation of solution, you are fixing a problem that you think you may have. You are not fixing the problem you have. Our management bias to fix before learning is so strong and so logical that it is hard to believe you don't know the problem, even though it is almost certain you won't know the problem. Amazingly, in most cases the problem you set out to solve is almost never the problem you end up learning and solving. Learning Teams provide better quality information about your workplace, and better information leads to solutions that are more effective.

Your current corrective action system probably fixes the wrong things well, with tremendous discipline and accuracy. Fixing the wrong things does not make your workplace better or safer. We simply must stop fixing what we think needs to be fixed and start understanding where the margins of effectiveness exist in the work in our organizations. Fixing what matters is better, requires less resource, and is much more satisfying and effective.

We can only correctly identify what needs work in our systems and processes by asking the people who really know how the work is completed. The days when we thought that the work the workers do on the shop floor is simple and easy are pretty much over; worker's today do very complex tasks in an incredibly complex environment. Managers are not smarter, nor more informed, then the workers who do the work. This operational complexity does not seem to be getting better; in fact, our operations appear to be moving to increasingly complex levels of process relationships.

Managers not knowing how work is done is not a criticism of managers not caring or not being smart enough. This is a criticism of managers not having the right information, the most accurate understanding of how work was completed. When managers push back, we have the opportunity to teach these managers a new way to learn.

We are in the midst of a change in how we perceive workers doing work. Where once we thought workers were somewhat uniformed users of the organization's processes and resources, we are now discovering the worker is a process owner and resource protector. Where we at one time hired workers for the back and brawn, we now are discovering a need to hire workers for their brains and agility. Our world is requiring we identify workers not as a problem to be fixed, but instead a resource to be harnessed for improvement and learning.

## *How we ask questions changes how workers answer questions*

This entire book is about changing the questions we as managers and leaders ask. We have not been asking good questions or the right questions. The old questions we ask only re-enforced the old problems we believed we had. We must ask better questions to get a better understanding, a better explanation, of how an event or accident transpired.

Could it be that simple?

The answer is both "yes" and "no." Yes, asking better questions really does produce much better answers; better answers absolutely do lead to better corrective actions; better corrective measures help make our workplaces safer and more reliable. For too long we have asked a "fixed" set of questions, questions we thought were correct, proper, and right. Questions that we honestly believed would make our world better and our workplaces safer. We are only beginning to understand how much damage our need to "stick just to the facts" has done to our ability to recognize the context of the event that existed for the worker doing the work. We have been led to believe that "why the worker did what he did" is better than "how the worker did what he did" as a method to investigate and learn about an event. We have even created standardized and rigid ways to do learning, ways that are much more concerned with formalizing our investigation process instead of a richer, deeper understanding or how a failure (or near failure) could have happened.

We did all of these things to make our organizations better. The problem is these formalized and restricted ways we are told to ask questions has limited the context information we are able to learn about our workplace. When we limit the questions we ask to just the facts, for example, we are also limiting our thinking about the event to just the things we can touch, see, and prove. Most operational decisions involve thinking, knowing, and feeling. Thinking, knowing, and feeling are not facts – yet these things actually happened – these important contextual factors have not been a part of our learning because these important contextual factors are not included in our questions.

Limiting how much we learn about an event or near event is not good for your organization, your workers, and importantly yourself. Knowing less about an event does not make you smarter or safer. Think of this idea based upon your own professional history. How many events have you reviewed where good workers seemingly have done unexplainable things on your shop floor? Normally dependable, always effective, historically successful workers fail in places and processes where they have not failed before. Do you really believe these worker actions cannot be explained?

It is not acceptable to say, "I will never understand how that guy made that decision that caused that outcome." This statement says nothing about the event and everything about your knowledge of the event. If you are not able to explain the event beyond something mysterious happened, you are not finished learning about the event. Remember, not knowing does not make you smarter. Not knowing is not a place where thinking can stop. Not knowing means you have to look deeper and harder at the event or near-event in order to know how something happened.

If you can't explain what happened, you cannot fix what happened. It is honestly that simple. You don't have to find an outside expert to tell you what happened; you don't need to have the best incident investigator to determine the chronology. You do have to ask the people who do the work, the people who had the event, to tell you the conditions needed for this failure to happen. Once you know what conditions were present, you can begin the process of restoring your workplace back to stable and safe performance.

The quickest way to change safety programs in our workplaces is to change how we define safety. The very act of moving the conceptual and actual definition of safety from an outcome to be achieved to a capacity to be managed will immediately change the way leaders and workers think about creating the capacity to be safe, safety. This change in thinking will absolutely lead to changing what questions we ask and what we look for within our organizations. Knowing that safety is a positive addition of capacity to your organization is much better than the traditional view around the absence of injury. This shift in thinking is important, significant, and an on-going conversation that seemingly will never end.

This entire discussion boils down to knowing more. We must take our learning to deeper and deeper levels within our systems and processes.

We must not stop at the first "cause" that makes sense for event. Knowing more about your systems in both the success mode and the failure mode, knowing more about how your workers actually perform work for your organization, knowing that the leadership view of work is much different from the worker's view of work are the necessities.

Knowing more about your organization as a place where systems meet people is what safety improvement for the new view is all about. We must go farther and dig deeper to learn more. The quickest way to shifting your reliability program is to improve the way you do operational learning. If you are better at learning, you will be better at being safe and reliable. Program delivery is not our biggest problem. Program execution has increased year after year, steadily in both quality and quantity. We most likely don't have to change the way we do safety, safety management systems and safety programs. We must change the way we think about managing safety.

If you are better at learning, you will be better at being safe and reliable.

Remember, really great organizations are constantly monitoring what is happening, not waiting for something to happen. Learning about problems early gives your organization the capacity to solve problems early, often before they are even problems. Knowing where problems are beginning to appear is a skill that few in your organization have. You probably don't have the access or skill to do it, but your workers do. Tapping into that skill is vital for your ability to assist your organization in learning and improving.

You don't have to change everything. For that matter, you don't have to fix everything. Most importantly, you can't fix everything. You have to change the way you think about everything your company or organization does in order to safely and reliably perform the work you do and to open up the potential to learn and improve. This shift in thinking is vital, the timing is right, and you are ready for this change.

All of this improvement begins by asking better questions.

# Index

Printed in the United States
by Baker & Taylor Publisher Services

Printed in the United States
by Baker & Taylor Publisher Services